CARBON FIBERS

CARBON FIBERS

Technology, Uses and Prospects

Edited by

The Plastics and Rubber Institute
London, England

Library of Congress Catalog Card Number: 86-5155
ISBN: 0-8155-1079-9
Printed in the United States

Published in the United States of America by
Noyes Publications
Mill Road, Park Ridge, New Jersey 07656

10 9 8 7 6 5 4 3 2 1

Library of Congress Cataloging-in-Publication Data

Carbon fibers.

Bibliography: p.
Includes index.
1. Carbon fibers--Congresses. I. Plastics and Rubber
Institute.
TA418.9.C6C29 1986 620.1'93 86-5155
ISBN 0-8155-1079-9

Foreword

Carbon fibers, their technology, uses and current prospects, are described in this book, based on the 3rd International Conference on Carbon Fibers, held in October 1985, and sponsored by The Plastics and Rubber Institute. Composites based on carbon fiber reinforcement (CFRPs) present designers with sets of properties and fabrication opportunities which offer very significant potential advantages. Progress towards realizing this potential is reviewed in the book in terms of the technology that has been developed for a range of products. It is also suggested, in addition to establishing markets for CFRP products, that carbon fibers have positively influenced the whole advanced composites sector. The pattern of market application development for CFRP is attributed to specific cost/product performance criteria.

The development of carbon fibers in the mid-1960s was an event of primary significance in the history of materials, representing a quantum advance in materials of low density which were both strong and stiff. In materials science terms it was a major advance. The importance of carbon fibers was that they offered a practically attainable form of material which could be produced in volume at potentially realistic cost. The composite materials based on carbon fibers defined a new frontier in the spectrum of engineering materials. An appreciable change in structural efficiency (in terms of performance/weight) was now attainable.

Why, therefore, in view of such apparent structural performance potential has the transition to engineering application been so cautious and the build-up rate of commercial markets so modest? Only in the last five years has a substantial market base been consolidated and worldwide investment committed to carbon fiber production capacity. The recent significant increase in supply now surely confirms the acceptance of carbon fiber as a viable industrial material. The catalytic influence of carbon fibers can justifiably be claimed to have been a major factor in establishing advanced composites as an essential part of the new materials technology, now widely accepted as a key requirement for future industrial development. In parallel with this technological progress a realistic commercial applications perspective has emerged.

Early attempts to use carbon fibers in products with commercial potential were wide ranging and often lacked sensible evaluation of performance benefits and acceptable cost criteria. The combination of exuberance to use the "wonder material," cost uncertainty, and the wholly inadequate CFRP technology resulted in a great deal of misapplication and failures both in commercial and technical terms. However, much of the valuable experience gained at that early stage of the CFRP learning curve has contributed greatly to establishing a basis for viable CFRP markets. Numerous applications for the use of carbon fibers today are to be found in the aerospace industry, vehicle manufacture, and the sports and leisure goods industry. An assortment of other interesting possibilities exists in fields as diverse as precision engineering, medicine and electronics, which could develop into special application areas of significance. The experience and development progress achieved in the early years of carbon fibers should provide a solid base for continuing vigorous growth of the applications market into a diverse range of products.

The information in the book is from *Carbon Fibres—Uses and Prospects,* proceedings of the 3rd International Conference, edited by The Plastics and Rubber Institute, London UK, October 1985.

The table of contents is organized in such a way as to serve as a subject index and provides easy access to the information contained in the book.

Advanced composition and production methods developed by Noyes Publications are employed to bring this durably bound book to you in a minimum of time. Special techniques are used to close the gap between "manuscript" and "completed book." In order to keep the price of the book to a reasonable level, it has been partially reproduced by photo-offset directly from the original report and the cost saving passed on to the reader. Due to this method of publishing, certain portions of the book may be less legible than desired.

NOTICE

Contents and Subject Index

PART 1
RAW MATERIALS—FIBER PRODUCTION AND DEVELOPMENTS

Part 1

Raw Materials—Fiber Production and Developments

IMPACT OF CARBON FIBERS ON FRP TECHNOLOGY

W. Paton

National Engineering Laboratory
East Kilbride, Glasgow

The composites based on carbon fibre reinforcement present designers with sets of properties and fabrication opportunities which offer very significant potential advantages. Progress towards realising this potential is reviewed in terms of the technology that has been developed for a range of products. It is also suggested, in addition to establishing markets for CFRP products, that carbon fibres have positively influenced the whole advanced composites sector. The pattern of application market development for CFRP is attributed to specific cost/product performance criteria.

INTRODUCTION

The development of carbon fibres in the mid 60s was an event of primary significance in the history of materials, representing a quantum advance in materials of low density which were both strong and stiff. In materials science terms it was a major advance to make such progress towards the theoretical ultimate offered by pure graphitic whiskers which were known to have extremely high values of Young's modulus (680 GPa) in preferred crystallographic directions. The importance of carbon fibres was that they offered a practically attainable form of material which could be produced in volume at potentially realistic cost. The composite materials (CFRP) based on carbon fibres defined a new frontier in the spectrum of engineering materials, with virtually a step change in structural efficiency (in terms of performance/weight). One interesting effect has been the response from the conventional structural metals - high strength low alloy steels (HSLA), aluminium/lithium alloys and titanium alloys. The relative magnitude of these improvements, for example the 25 per cent modulus increase in Al/Li, merely underlines the huge advance which carbon fibres represents.

Why, therefore, in view of such apparent structural performance potential has the transition to engineering application been so cautious and the build-up rate of commercial markets so modest? Only in the last 5 years has a substantial market base been consolidated and world-wide investment committed to carbon fibre production capacity. The recent significant increase in supply now surely confirms the acceptance of carbon fibre as a viable industrial material. Some of the answers lie in the understandable reluctance of designers to use such unproven materials, the paucity of concepts and methodology for fully utilising the unique sets of properties which composites offer and the inadequate technology for translating these properties into products. In the intervening 15 years great progress has been made on each of these fronts and the catalytic influence of carbon fibres can justifiably be claimed to have been a major factor in establishing advanced composites as an essential part of the new materials technology, now widely accepted as a key requirement for future industrial development. In parallel with this technological progress a realistic commercial applications perspective has emerged, exciting in many respects but certainly less spectacular than some of the scenarios envisaged in the early days. This paper looks at some of the main features and highlights in the 15-year development of carbon fibre composites in industrial technology and considers the impact they have had on the broader FRP technology that has grown around it.

FIBRES AND RESINS

Early developments in CFRP depended on first generation fibre types with variable, under-developed and non-optimised properties combined with a limited range of available resin systems usually developed for other purposes. In addition, surface treatments crucial to the interfacial characteristics were crude and the resultant composites often characterised by high scatter in the mechanical properties, poor predictability and serious deficiencies in specific properties such as impact toughness, inter-laminar shear strength and environmental resistance. It was hardly surprising that these shortcomings, along with an inadequate specification basis for composite systems, were seen by serious designers as a major impediment to immediate use. The great progress that has been made since that early development phase toward real 'engineering materials' status can be attributed to several factors.

- Improved production technology for fibres yielding well-defined mechanical properties with acceptable variability, in line with design requirements (eg higher failure strains) and sensibly optimising cost/performance criteria.

- Customised resin systems which are more compatible with the reinforcement in thermo-mechanical terms and provide the essential levels of environmental resistance (hot-wet strength) and fracture toughness.

- Major materials science input which has progressively established a reliable basis for understanding composite micromechanics, failure modes and material response in a wide range of mechanical, chemical and physical conditions.

- User feedback of the improved design, performance and fabrication requirements leading to better definition of material development and objectives.

As a consequence of the sustained development and supporting research on these fronts, standard CFRP composite systems are now available which provide an acceptable basis for structural design. Fibres with higher strain to failure of 1.5-1.8 per cent have been developed which greatly improve impact strength. Typically these fibres are in the 230-260 GPa modulus range with tensile strengths of 3500-4500 MPa. Higher modulus fibres (345-400 GPa) with relatively low failure strains (.5-1 per cent) are available for more specialist applications, often where high mechanical and thermal stability is needed. The trend in fibre development has been towards quality and low properties scatter with market-led compromise between modulus, strength and production costs. Efficient translation of these properties into resin based composites produces a group of CFRP materials with an impressive range of properties. Typical values for basic unidirectional and bidirectional materials are given in Table I.

Epoxy resins remain the most widely used matrix material for general CFRP application and major advances have been made with high elongation, toughness, hot-wet strength and high temperature capability for sustained performance above 130°C. Special epoxies are being challenged by bismaleimides, polyamide/imide and phenolics. Thermoplastics are now receiving considerable attention because of possible fabrication advantages and improved toughness. Polyphenylene sulphide (PPS), polyether-sulfone (PES) and polyether-ether-ketone (PEEK) are typical of new types with better toughness and resistance to wear, aggressive environments and high temperature conditions.

PROCESSING AND FABRICATION TECHNOLOGY

Pre-impregnated sheets and tapes are now produced with tight specification of fibre content, ply thickness and handling characteristics by automated processes with a high level of on-line control. These have become the basic building units of high performance CFRP design, bridging the large gap between the complexity and uncertainty of material formulation and the essential requirement of the designer for properties data which are both well-defined and reproducible. There is little doubt that CFRP has accelerated the development of pre-impregnation technology both in terms of precision and quality of product (voidage, homogeneity and thickness control) and fabrication characteristics (formability, tack, consolidation control). Machine compatibility for automated lay-down systems is an increasingly important driving force as high productivity manufacturing routes are sought.

Textile techniques have been progressively used to produce a wide choice of woven fabrics which have good forming/draping characteristics and can be easily handled in manufacturing processes. A key feature of these forms is the facility with which fibre configurations can be arranged specifically to match the design loading patterns. The improved understanding of the mechanics and failure behaviour of composites with woven reinforcements has underpinned the development of a range of important intermediate products with useful design features.

- Carbon/aramid/glass hybrids with wide ranging properties/cost options.

- Precise control of anisotropic ratios.

- Enhanced properties such as impact toughness and thermal stability, often with 3D reinforcement.

The adaptation of the braiding process illustrates the scope for mechanised production of flat tapes or tubular forms with unique formability and the angular reinforcements essential for high shear properties (Figure 1). Textile based products clearly have the potential for high volume production either as a feedstock to a specific moulding process or, for example, in combination with the injection of fast curing resins as a direct manufacturing process. The use of automated methods for pre-impregnation cutting, ply nesting and pre-forming can greatly reduce the cost of manufacture, provide better quality and ensure reproducibility. The multi-robot cell developed at NEL for pre-forming pliable materials is one of several developments which demonstrate a possible basis for volume manufacture (Figure 2). With interface to computer-based design stages and product inspection, fully integrated production systems can be envisaged.

APPLICATION DEVELOPMENT

Early attempts to use carbon fibres in products with commercial potential were wide ranging and often lacked sensible evaluation of performance benefits and acceptable cost criteria. The combination of exuberance to use the 'wonder material', cost uncertainty and the wholly inadequate CFRP technology resulted in a great deal of misapplication and failures both in commercial and technical terms. However, much of the valuable experience gained at that early stage of the CFRP learning curve has contributed greatly to establishing a basis for a viable CFRP market. Following that initial transient phase there was undoubtedly a better understanding of the technology gradient to be ascended and a more realistic perspective of likely areas of advantageous application. It was clear that CFRP would remain a relatively expensive material and that the cost premium incurred in its use would have to be justified in terms of overall product advantages derived from the special properties and characteristics of the material. Since weight reduction was the most obvious primary benefit, components and structural parts associated with aircraft and space vehicles soon became a major focus of investigation. Weight savings could be directly related to operational and payload advantages which often had quantifiable and acceptable economics. Similarly, sports and leisure equipment emerged as a promising market sector. The utility of high specific mechanical properties in dynamic design pointed the way to a variety of high speed machine components ranging from textile machines to centrifuges. There was also a realisation that CFRP was only one possibility within a much broader family of advanced FRP composites, a factor which has stimulated a strong competitive response to carbon from other types of reinforcement.

Aerospace

Aerospace has been a primary influence in the progress of CFRP and now constitutes the largest market sector. It is estimated to account for 60-70 per cent of the current world-wide output of 2000-2500 tonnes of fibre and seems likely to be the source of sustained growth over the next decade of at least 10 per cent per annum. A feature of this expansion will be greater usage in helicopters and civil aircraft, particularly in Europe. In addition to the already significant consumption in military aircraft, satellites, other space and defence requirements are also likely to increase. Carbon fibre/epoxy systems have accounted for the greatest proportion of applications although there have been important developments in carbon fibre/carbon composites relevant to aircraft brakes, nozzles and other high temperature components. It is beyond the scope of this paper to consider the full extent to which CFRP has become accepted as a first choice material in many structural and functionally critical parts. Figures 3 and 4 illustrate some of the areas in which advanced composites are considered to be competitive with metals. Carbon fibres are preferred in those stiffness-critical components where design specifications on, for example, deflection limits, buckling and dynamic response cannot be met without them. Important recent advances include major use of CFRP for control surfaces (for example, spoilers and ailerons) and tailplane assemblies for the Airbus 320 and the Boeing 737 aircraft. Further progress towards primary structural utilisation for fuselage, wings and helicopter airframes will depend very much on CFRP. The technical potential has been demonstrated in the Lear Fan 2100 (CFRP fuselage and wing structures), the Grumman X29 (advanced composite wing design) and the British Aerospace advanced fighter demonstrator.

The achievement of the aerospace industry over 20 years has been impressive; CFRP has become an accepted part of the design and manufacturing technology, cost effective benefits have been established, and a substantial commercial market for CFRP has been built. This has only happened because of a high level of commitment and investment in research, development and demonstration. There are some important guidelines to successful applications to be drawn from aerospace experience.

1 A comprehensive design capability has been established based on a detailed understanding of composite characteristics and safe design limits. This enabled material design to be optimised in relation to structural requirements and facilitated the adoption of new construction concepts with greater parts integration.

2 Validated design data relevant to the service environment have been generated in extensive testing programmes. Emphasis on the testing of real parts under, for example, fatigue conditions, has increased confidence in performance prediction to an acceptable level.

3 Realistic design objectives have been set taking account of the total engineering requirement. This has resulted in lower structural weight savings, typically from 20 per cent (wing assemblies) to 45 per cent (fin and tailplane structures) than suggested from simple materials data. Complex loading combinations, the need to retain some metal, for example at joints, and the relatively low design strain limits account for this.

4 Manufacturing technology has been developed which is appropriate to the quality, standards, reproducibility and volume requirements of the industry. The acceptance of new concepts, often involving the combination of different composite types and forms and the integration of several assembly stages into one operation, has been an important factor in being cost effective despite high material cost.

5 The importance of full evaluation of the cost implications of manufacture, assembly and in-service operations, has been accepted as a prerequisite for successful innovation.

Vehicles

In contrast with the aerospace sector, where CFRP design, manufacturing and application expertise has developed with a degree of coordination towards definable objectives with viable economics, technical progress in vehicles has been fragmented and commercial development more uncertain. There is little doubt that there are similar incentives to use FRP material, notably for weight reduction and parts integration, but the volume requirements and cost criteria have proved to be major obstacles, particularly in the case of CFRP. Even with the general drive to lightweight, fuel efficient vehicles, it has been difficult to justify the high cost of CFRP on cost/performance benefits. Nevertheless, carbon fibres have greatly assisted in promoting an awareness within the vehicle industry of the scope for using FRP composites in load bearing parts.

The technical feasibility of CFRP was demonstrated by Ford (US) in their lightweight vehicle project by successfully replacing steel parts such as doors, bonnets, wheels, drive shafts, structural brackets and frames with a resulting weight reduction from 1820 to 1250 kg. Engine components (conrods, pistons and rocker arms) have been demonstrated in the novel

Polimotor and numerous CFRP constructions in racing car bodies have confirmed the view that CFRP can cope with stressed applications in the demanding vehicle service environment. Although technically CFRP would be the preferred choice in many designs, the potential application window narrows enormously when cost is considered even on optimistic projections of fibre prices as low as £15/kg. It follows that CFRP will only be used on a highly selective basis where stiffness or other special features are essential. Accepting this role, there is nevertheless a substantial market potential for CFRP as FRP applications generally expand. The prospect of 1000 tonne increments in demand for every 100 grammes of carbon used on a typical medium volume passenger car range is a powerful incentive.

Possible applications have been identified in the suspension system, drive train and engine areas. CFRP triggered several developments in leaf springs in the early 1970s which showed considerable promise for low weight and fatigue tolerance. Progress from concept to product development towards the commercial market for light vans and trucks illustrates dramatically the cost barrier to carbon fibre usage. GRP/CFRP hybrid designs soon replaced all-carbon designs and with extensive design, materials, and process development, a 100 per cent GRP leaf spring became the commercial reality. Helical springs presented an even greater design challenge with the need for a high level of stiffness and fatigue strength in torsional shear. NEL has shown that CFRP can out-perform other composites in this mode and offer major performance/weight reduction compared with spring steel. Commercial development of this type of substitutional component will be critically linked to fibre price. The feature that has been highlighted in much of the work on conventional springs is the eminent suitability of FRP as an elastic energy absorbing medium. Thus, combined with other characteristics such as damping, property tailoring and ease of forming novel geometries, there appear to be many possibilities in 'vibration control' devices. The sulcated principle (Figure 5) developed at NEL may represent the direction of future innovation.

Automotive suspension systems perform complex structural functions and have a major influence on handling characteristics and ride quality. New concepts using FRP have received considerable attention in Europe with promising results, indicating considerable weight reduction, good fatigue performance and worthwhile gains in noise/vibration isolation. The technical feasibility of FRP in semi-trailing arms was established in glass hybrid component design by SEP/Peugeot. Carbon fibre was necessary to meet the total compliance specification within the packaging constraints. The situation is similar in some ways to the technically simpler case of the propeller shaft where optimisation of material design in relation to each identified structural function leads to a hybrid composite component with CFRP providing the flexural stiffness essential to dynamic specification. The propeller shaft is a good example of how composites can effectively penetrate existing markets by reducing assembly costs for the system as well as offering excellent performance characteristics. As in other cases the full utilisation of the potential for innovative designs leading to new types of components is likely to be the key to a commercial breakthrough.

Sports and Leisure

CFRP was received with great enthusiasm by designers of sports equipment because it offered many direct and indirect performance features which were marketable. It was also easier to assimilate in an industry familiar with diverse material combinations, receptive to innovative design and not constrained by the rigorous structural standards of the aerospace industry. From early applications in rackets (Figure 6), golf shafts, skis and various marine craft, a significant commercial outlet for carbon fibres has been built up. This has contributed greatly to the general growth of composites in quality leisure products with structural performance requirements. Added advantages of novel geometries, material combinations and a range of secondary properties which are of value in designing to criteria which may be partly subjective have proven to be of great benefit in this type of product design. Sophisticated production technology has also been developed in response to demands for volume and quality. The process developed by Dunlop for rackets is an outstanding example of how a product-led innovation can advance the technology, market potential and competitiveness. The basic idea of injection moulding carbon fibre/nylon hollow sections around fusible mandrels may find more widespread use.

Miscellaneous

The three sectors referred to above present reasonably well defined opportunities and are expected to grow. An assortment of other interesting possibilities exist in fields as diverse as precision engineering, medicine and electronics which could develop into specialist application areas of significance. The common feature that links many of them is the utilisation of material properties which may be individually or, in combination, unique to CFRP. This can be the basis for step changes in functional performance or even novel products. Some examples are given below of frequently recurring non-structural characteristics which, often together with the latter, make CFRP a primary design choice.

Property	Applications
Low thermal expansion	Satellite antennae and support structure Large telescopes, optical benches Waveguides for stable, high frequency (GHz) Precision measurement frames (Figure 7)
Bio-compatibility	Implants, prosthetics, tendon/ligament repair
Electrical conductivity	Novel tooling Casings and bases for electronic equipment (EMI and RF shielding) Brushes
Electro-magnetic	Large generator retaining rings Radiological equipment
Chemical inertness	Valves, seals and pump components in process plant
Wear/lubricity	Self-lubricating components in pumps and motors Precision machine components (particularly with injection-moulded thermoplastics)
Dynamic/vibration damping	Musical instruments Loudspeakers on Hi-Fi pick-up arms Robot arms

CONCLUDING REMARKS

The early commercial bridgeheads that carbon fibres established in the aerospace and leisure goods sectors have flourished into healthy markets and contributed significantly to the consolidation of CFRP as a growth technology. In both cases the cost premiums could be justified by quantifiable benefits in performance and marketability. Innovation in other areas has often been inhibited by cost, technical, and manufacturing factors. It is now appreciated that extended timescales may be involved in successfully introducing completely new materials and that early expectations of instant application success were often over-optimistic. The experience and development progress achieved in the first 15 years of carbon fibres should provide a solid base for continuing vigorous growth of the applications market into a more diverse range of products.

ACKNOWLEDGEMENTS

This paper is presented by permission of the Director, National Engineering Laboratory, Department of Trade and Industry. It is Crown copyright.

TABLE I Some typical mechanical properties (undirectional) of CFRP and other advanced composites

Composite	Tensile Strength GPa	Tensile Modulus GPa	Specific Gravity	Inter-laminar Shear Strength GPa
CFRP (standard)	1.52	131	1.58	0.1
CFRP (1.5% strain)	1.9	134	1.60	0.1
CFRP (1.8% strain)	2.6	138	1.61	0.1
CFRP (high modulus)	0.78	240	1.8	0.035
Aramid/epoxy	1.37	46	1.45	0.05
E-Glass/epoxy	1.14	41	1.9	0.08
S-Glass/epoxy	1.7	52	2.02	0.08
Aluminium alloy	0.43	70	2.7	-

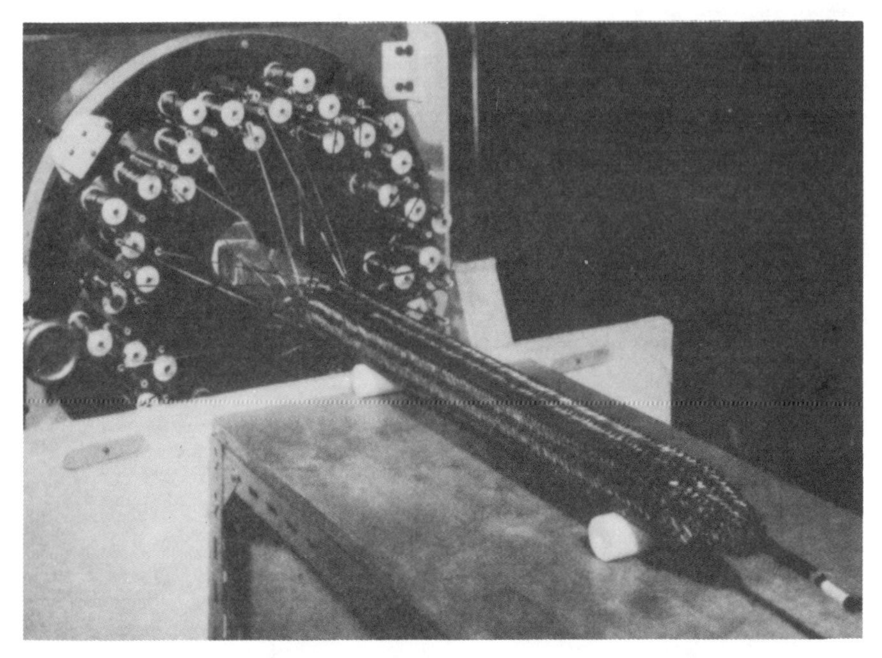

Figure 1 The braiding of carbon fibres into structural components

Figure 2 Multi-robot system for the build up of pre-impregnated materials

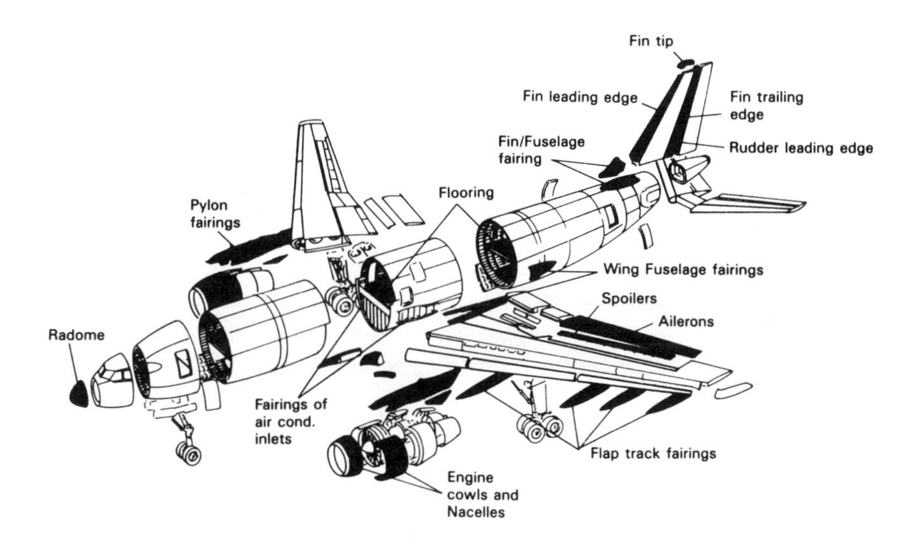

Figure 3 Areas of application of advanced composites on civil aircraft

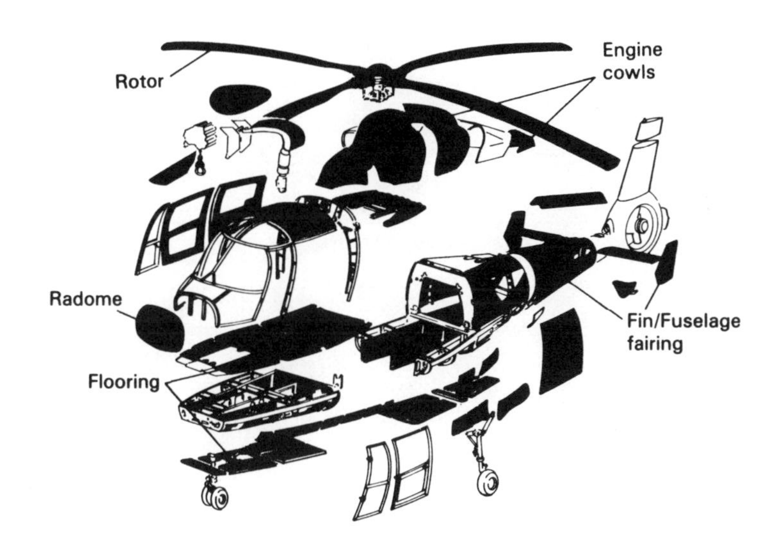

Figure 4 Use of advanced composites in helicopter construction

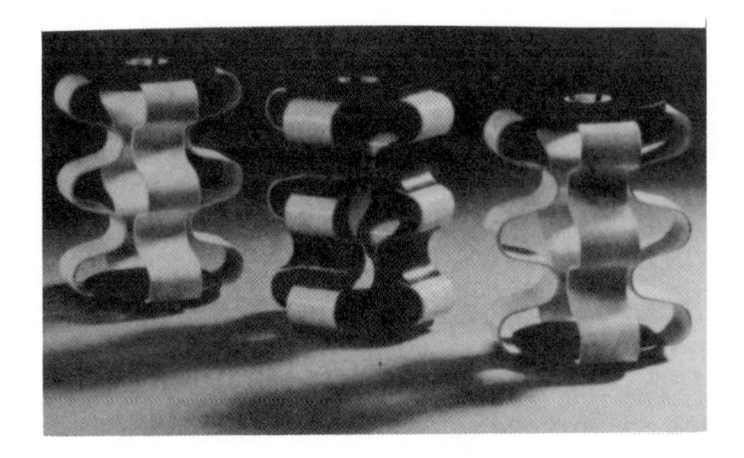

Figure 5 Sulcated principle of advanced composite spring

Figure 6 CFRP reinforcement of squash rackets

Figure 7 CFRP structures used for the calibration of coordinate measuring markings

MANUFACTURING ADVANCES TOWARDS HIGH STRAIN FIBERS

Yoshio Izuka, Toshiaki Norita, Tadashi Nishimura

Toray Industries, Inc.
2-2 Nihonbashi-Muromachi
Chuo-Ku, Tokyo, Japan

Kiichi Fujisawa

Toray Europe Ltd.
35/38 Portman Square
London W1H OBS, England

1. Introduction.

Toray has been commercially producing carbon fibers for more than decade. Through the period, TORAYCA T300 has rapidly grown in terms of quantity and also application, due to stable quality and large quantity of material data accumulation.

The applications of high performance carbon fiber, which initially found their usages in premium sporting goods, have been rapidly expanding in aerospace fields as a metal substitute for secondary and primary structures.

To fulfill requirements to carbon fiber composite as aircraft structure, several new types of carbon fiber have been developed. In this paper, the development activities of high strength fibers and high modulus fibers, and their composite properties are reported.

2. Required properties for aircraft structures.

The failure strain of T300 composite under various conditions is shown in Figure 1.[1] The failure strain of T300 composite was higher than 0.8% in case of without hole specimen. However, it showed 0.3-0.4% strain under condition of with hole or after impact damage incurred. From those data, design allowable strain for CFRP at present, to our understanding, has been fixed at 0.3-0.4%.

Figure 2[1] shows potential weight saving of T300 composite of two different laminate patterns $(0_2/\pm45)_s$ and $(0/\pm45/90)_s$ over Al-alloy(7075-T6).

In stiffness design, weight saving is achieved by higher specific modulus of CFRP over Al-alloy. Even with T300 composite, it can save weight of components by 20-50%. To gain more weight reduction higher modulus of composite is required. It has been known that the impact resistance can be improved with increase of the ratio of ±45 layer in lamination pattern. With increasing of the ratio of ±45 layer, modulus of the composite becomes lower, or the composite becomes thicker and heavier if the modulus is kept to the same level. Higher modulus of fiber can provide a composite with same thickness, weight and modulus with more impact resistance.

In strength design,however,current design allowable strain at 0.3-0.4% for T300 composite results in little weight saving. To obtain more weight saving,it is essential to develop carbon fiber and composite with larger allowable strain.

3. Factors which determine fiber strength and modulus.

If carbon fiber has complete graphite structure, the theoretical strength and modulus of carbon fiber were reported 180 GPa and 1020 GPa respectively.[2]

Actual value of modulus is 50% of theortical value with PAN carbon fiber and 80% with pitch carbon fiber.This may suggest that attainable modulus depends on the structure of carbon fiber. On the other hand,actual value of strength obtained is only 3-5 GPa with PAN fiber and 20 GPa even with graphite wisker. This suggests that strength of fiber is affected by defects of fiber rather than its structure.

As shown in Figure 3, carbon fiber strength varies with guage length of test and also cleanness of polymer dope and spinning atomosphere.[2] By observing fracture surface with much care, some defects can be detected and these defects can be thought the starting point of fracture. There are many types of defects and typical defects are voids and extraneous substance.

Figures 4 and 5 show relation between size of defect and fiber strength.[3]

4. Developments of fiber with high strength and high modulus.

As mentioned above,strength of carbon fiber is mainly dominated by the amount of defects in fibers. By setting up the entire series of manufacturing conditions to reduce any kind of fiber defects,the strength of fibers can be increased from 3000 MPa to almost 6000 MPa.

On the other hand,by improving the perfectness and the orientation of fine structure and crystallites of carbon fiber,the high modulus fiber of 500 GPa was developed. And the development activities were done to improve strength of high modulus fiber.

All the TORAYCA carbon fiber products are summarized in Figure 6.

5. Properties of fibers and their composites.

5.1. Fiber properties.

The properties of TORAYCA fiber family are summarized in Table 1.

* T300 is a base product of TORAY high strength fiber.
* T400 is high strength fiber.
* T800 is a high strength fiber with intermediate modulus.
* M30 is a intermediate modulus fiber.
* M40 and M50 are high modulus fiber.

5.2. Composite properties.

Various properties of T800 composites were evaluated. The results are summarized in Table 2 and 3.[4] compared with T300 composite and Al-alloy 7075-T6. The improved properties by T800 are marked with ⟶ , and the non-improved with ⟶ in Table 3.

The most significant advantage is the improvement of modulus. T800 unidirectional composite shows 20% higher modulus in fiber direction than T300 composite.

This advantage is also fully reflected in isotropic lamination pattern. T800 composite shows 50% higher specific modulus than the Al-li [5] alloy as illustrated in Figure 7. This improvement will provide designing engineers more extensive freedom as structural design for the stiffness of critical component. For example more ±45 laminates can be used, keeping the same stiffness as T300 composite.

The another advantage is the improvement of longitudinal tensile strength, which is the essential property for the actual design, as high as 3000 MPa, as shown in Figure 8. This improvement also provided higher fatigue resistance as illustrated in Figure 9.

T800 isotropic composite also showed higher tensile strength without and with hole as well as improved fatigue property than Al-alloy, as illustrated in Figure 10. Comparison of through-penetration impact of T300 and T800 is shown in Figure 11.

Since the longitudinal tensile strength is greatly affected by the matrix resin as shown in Figure 12, the selection of proper resin system or higher elongation resin is very important to obtain higher composite strength or higher translation of fiber strength into composite.

In spite of the improvement of tensile strength and through-penetration impact load,compressive properties are not improved. Compressive strength of T800 composite and Al-alloy were evaluated using the specimens without and with hole at elevated temperature under dry and wet conditions as shown in Figure 13. The compressive strength of the composite is remarkably affected by the existence of hole in the specimen. This is presumed to be due to the difference of material response to stress concentration around the hole. Delamination can be another reason for the drop of compressive strength of composite. To solve these problems,the development of tough resin and the implovement of interface property between matrix resin and carbon fiber might be very important.

6. **Summary**.

To fulfill the requirements to carbon fiber composite for aircraft structures,a variety of high strength carbon fibers such as T800 and T400 were developed by the set up of the entire series of manufacturing conditions so as to reduce any kind of fiber structual defects on the the basis of T300 technology.Also new type of high modulus fibers were added to existing product mixes.

The T800 composite shows greatly improved properties dominated by fiber, such as 20% higher tensile and compressive modulus and 60-80% higher tensile strength compared with T300. Through-penetration impact load is also increased to two times of T300 composite. Compressive and traverse tensile strength of T800 composite do not reflect the improvement of fiber properties. The development of tough resin and the improvement of interface properties between carbon fiber and matrix resin are required.

Besides the improvement of compressive property of composite,manufacturing cost saving,inspection and evaluation technique to ensure the reliability and design technology to use the unisotropic properties of composite will be the most important items to be worked on for the wider application of carbon fiber composite in future.

7. References.

1) Boeing Presentation Document.

2) W.N.Reynolds et al., Philoso. Trans.R.Soc.of London,P451,A29 (1981)

3) K.Noguchi et al., Carbon'84 (Int.Carbon Conf.) Preprint

4) J.Matsui,T.Norita et al., Material Sci.Monogr.29 (1985)

5) W.E.Quist and G.H.Narayanan, 28th International SAMPE Sympo.,Preprint P 374 (1983)

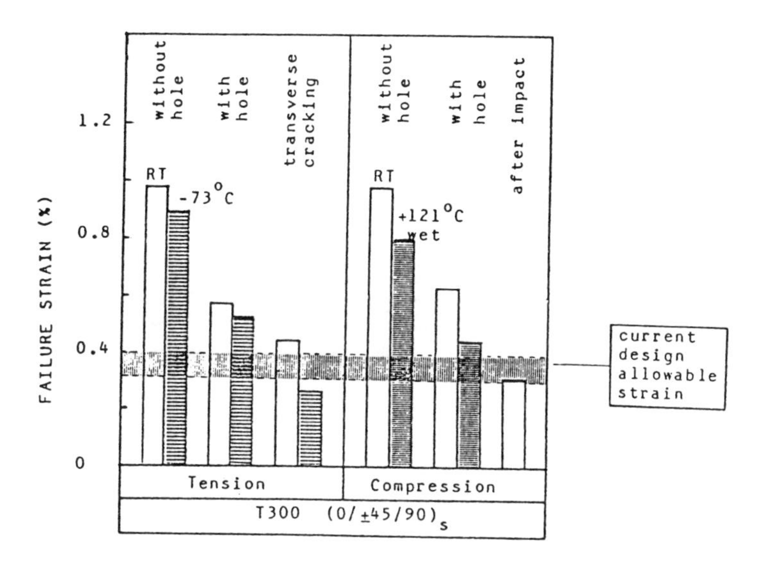

Figure 1. Failure strain of T300 composite under various conditions

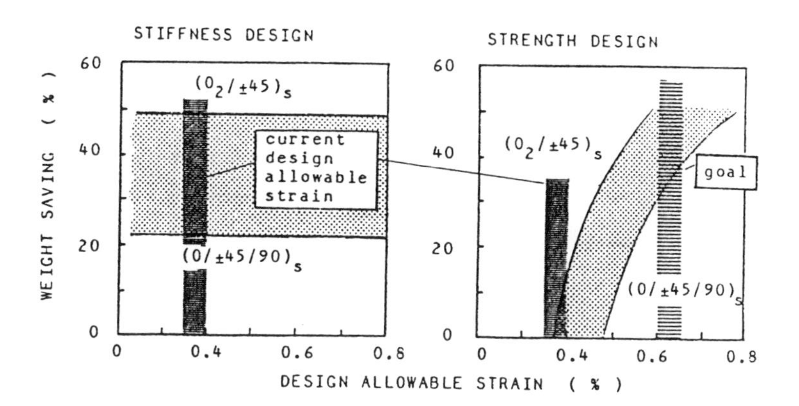

Figure 2. Weight saving and design allowable strain under two design concepts, T300 composite versus 7075-T6

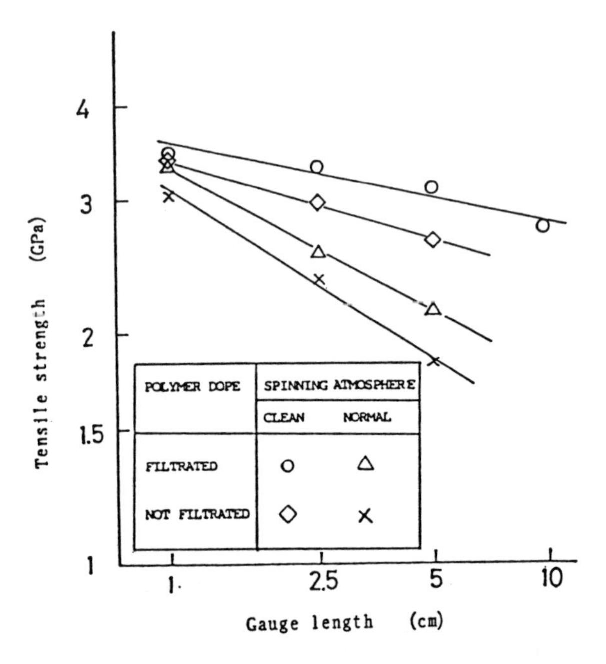

Figure 3. Effect of cleanness of polymer on tensile strength of carbon fiber

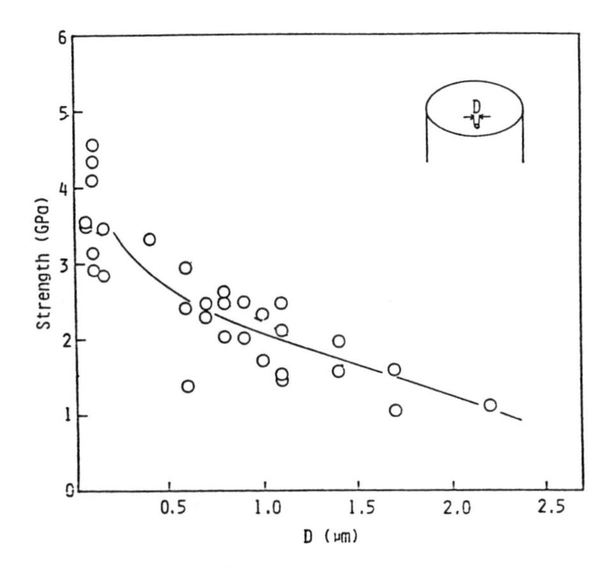

Figure 4. Dependence of strength on void size

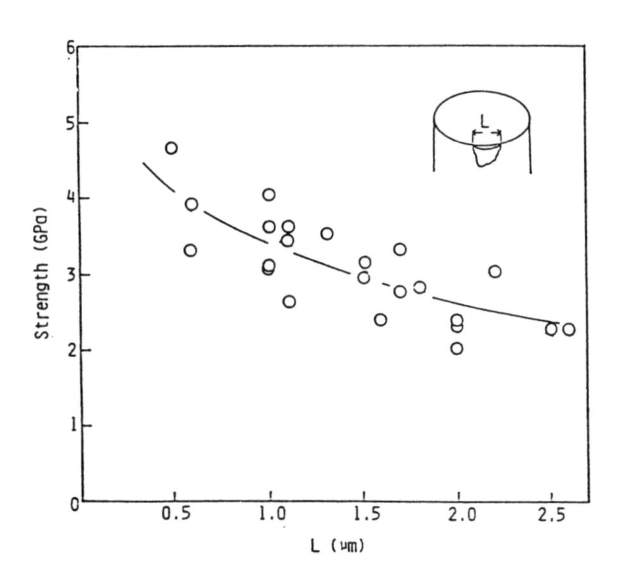

Figure 5. Dependence of strength on size of extraneous substance

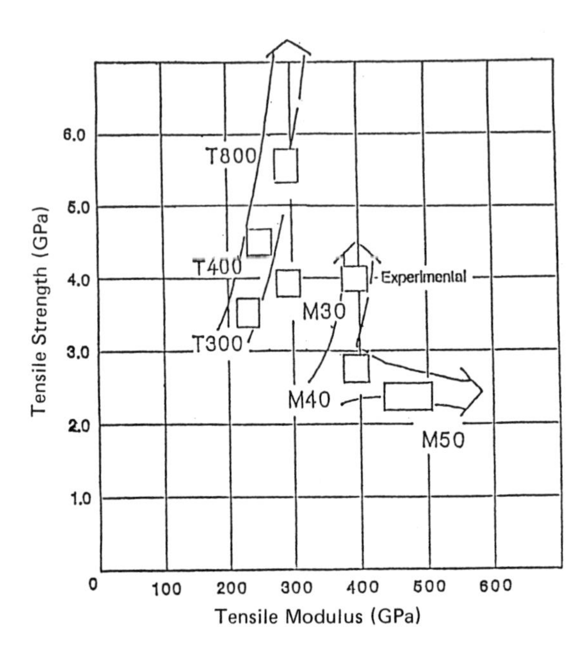

Figure 6. Development of TORAYCA carbon fibers.

Table 1. Properties of TORAYCA carbon fiber

Type	Filaments	Tensile Strength		Tensile Modulus of Elasticity		Elongation	Yield	Density
		MPa	kgf/mm²	GPa	kgf/mm²	%	g/1000m (tex)	g/cm³
T300	1,000	3,528	360	235	24,000	1.5	66	1.76
	3,000	3,528	360	230	23,500	1.5	198	1.75
	6,000	3,528	360	230	23,500	1.5	396	1.76
	12,000	3,430	350	230	23,500	1.4	800	1.77
T400	3,000	4,508	460	250	25,500	1.8	198	1.80
	6,000	4,508	460	250	25,500	1.8	396	1.80
T800	12,000	5,586	570	294	30,000	1.9	455	1.80
M30	3,000	3,920	400	294	30,000	1.3	160	1.72
	6,000	3,920	400	294	30,000	1.3	320	1.72
M40	1,000	2,744	280	392	40,000	0.6	61	1.81
	3,000	2,744	280	392	40,000	0.6	182	1.81
	6,000	2,646	270	392	40,000	0.6	364	1.81
M50	1,000	2,450	250	490	50,000	0.5	60	1.91
	3,000	2,352	240	490	50,000	0.4	180	1.91
	6,000	2,352	240	451	46,000	0.5	360	1.88

*TORAYCA fibre properties are measured by the test methods described in TORAYCA technical data sheets TY-020A, TY-030A-01, 02, 03

Table 2. Improvement of Modulus

Matrix: #3620, Vf=60%

Item	Direction	T300	T800
Tensile Modulus Et (GPa)	0°	133	163
	90°	8.8	8.8
Compression Modulus Ec (GPa)	0°	121	147
	90°	8.8	8.8

Table 3. Properties of high strength CFRP compared with Al-alloy

Material Fibre type			Composite* T800	Composite* T300	Alu 7075	Remark
Tensile Strength (MPa)	(0)		2940	1700	--	↗
	$(0/\pm45/90)_s$	without hole	820	554	540	↗
		with hole	660	323	460	↗
	(90)		60	60	--	→
Compressive Strength (MPa)	(0)		1570	1560	--	→
	$(0/\pm45/90)_s$	without hole	660	660	440	→
		with hole	350	350	440	→
	(90)		290	290	--	→
Impact	Through penetration load (N)** $(+45/90/-45/0/+45/90)_s$		440	190	485	↗
	Compressive strength after impact (MPa) $(0/\pm45/90)_{4s}$					
	Energy (J/cm)	0	490	490	440***	
		67	150	160	440	→
		penetration	95	100	370	

* Matrix resin #3620 ; Fibre content 60% volume
** Panel thickness 1.64 to 1.70mm
*** 7075 panel thickness 5.0mm

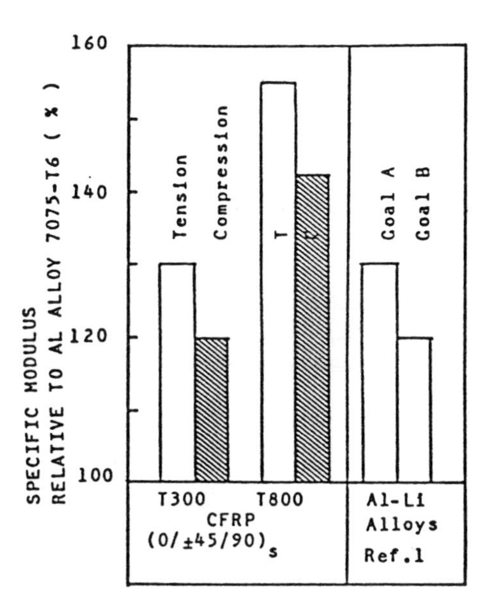

Figure 7. Comparison of specific modulus

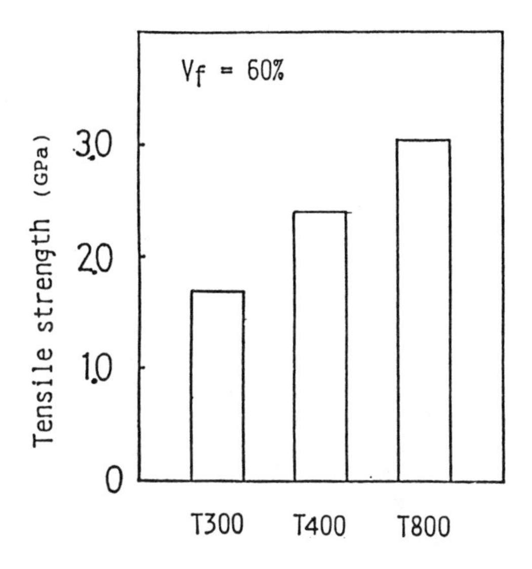

Figure 8. Comparison of longitudinal tensile strength (Resin #3620)

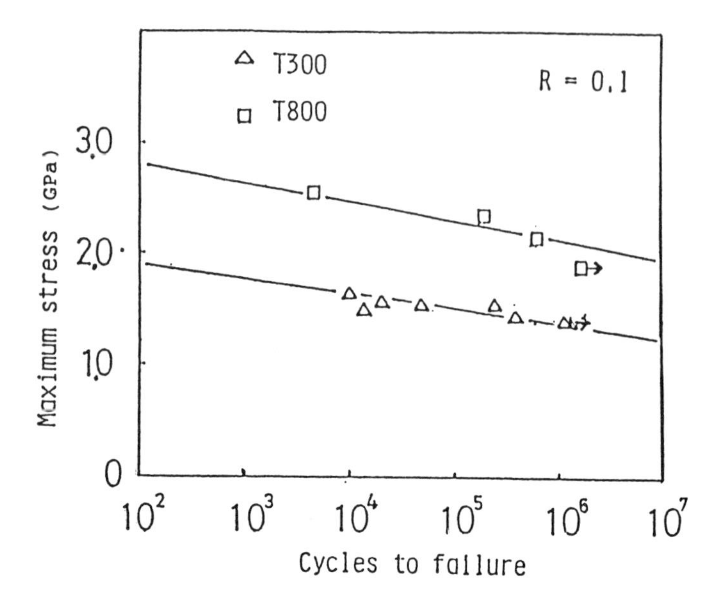

Figure 9. Comparison of fatigue resistance (Resin: #3620)

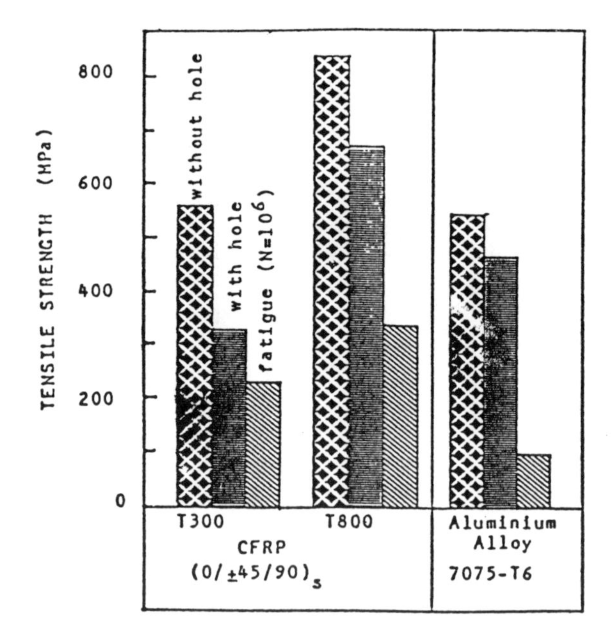

Figure 10. Comparison of tensile strength

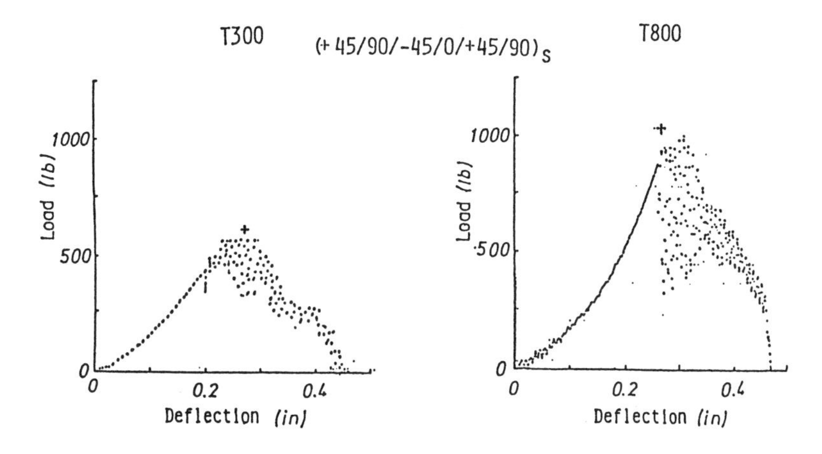

Figure 11. Comparison of through-penetraion impact (Resin: #3620)

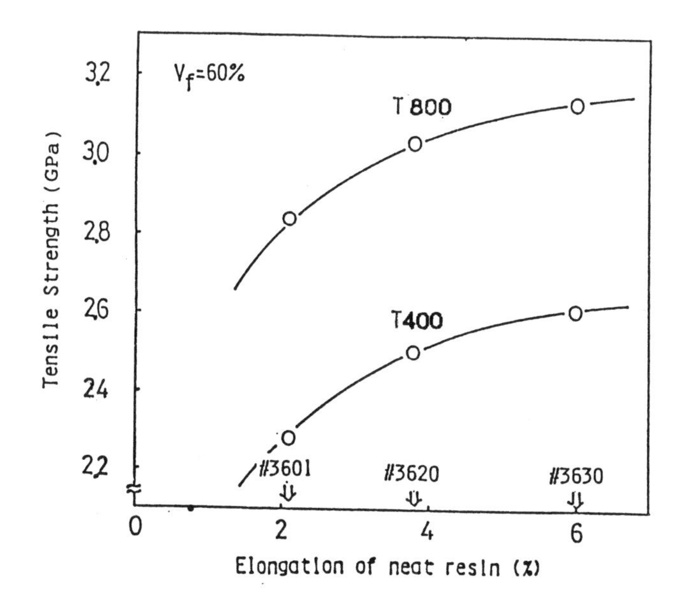

Figure 12. Effect of matrix resin on longitudinal tensile strength.

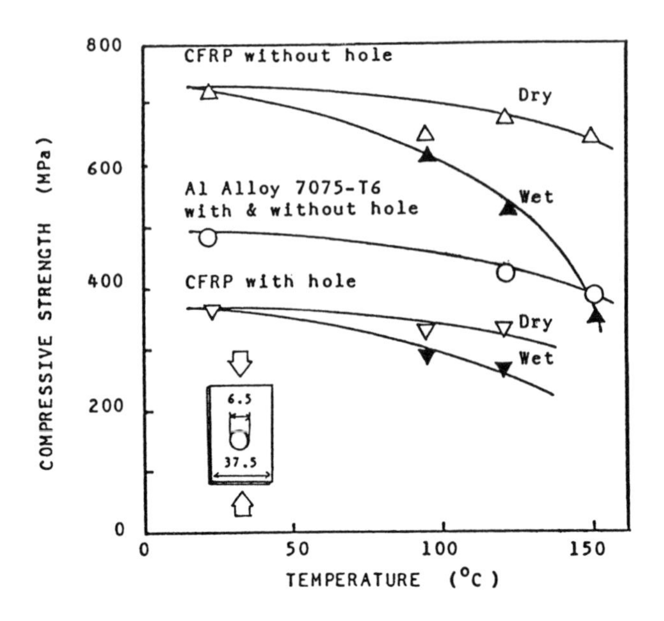

Figure 13. Comparison of compressive strength
(Resin: #3620)

HIGH PERFORMANCE FIBERS: THE DEVELOPMENT OF IMPROVED PROPERTIES IN THE HYSOL-GRAFIL RANGE OF PRODUCTS

C.F. Holleyman

Courtaulds Research
Courtaulds P.L.C., Coventry

Recent R & D effort to enhance the properties of the Hysol-Grafil range of products has been directed at reducing particulate contamination within the process and improving the morphology of the final product. Improved filtration has offered over a 20% improvement in strengths over the last 2 years. In addition the development of a new process has led to the launch of a new range of products with significantly improved properties.

INTRODUCTION

In the last 8 years worldwide carbon fibre market demand has increased rapidly by nearly 15 times, from a level of 255 tonnes/annum in 1977 to the current 1985 level of about 3,800 tonnes/annum (Table I). This attractive growth is expected to continue on into the 1990's following a traditional 'S' substitution development.

Across the period there have been a number of outstanding features of the development, each of which has significantly influenced the direction and pattern of growth in end-use development. The key factors here have been the dramatic reduction in production costs and selling prices, but of possibly more importance, the industry has offered very tangible improvements in product quality and properties.

In the early days a high proportion of the final product was sold into the sports goods markets. However, against the background of improving properties and growing confidence in the product, the last few years have been marked by a materials and technology revolution in the aerospace industry. The net result has been a very rapid increase in the use of high performance carbon composites in this area. The main driving forces behind this development have been the significant weight and cost benefits offered by composites over conventional structural materials.

The first stage of substitution of carbon fibre composites into the aerospace market involved assessment of the material in non-critical areas of aircraft and satellite structures. However, now with a database and growing confidence in the new material the trend has been towards increasing use in more critical areas such as carbon-carbon brakes and primary aircraft structures.

Paralleled by this move there has been increasing pressure from aircraft designers calling for enhanced composite performance which has prompted major research into improving both fibre and resin properties.

In response to the market the last few years have seen very marked improvements in the properties of the "Grafil" range of products. The extent of this is illustrated by Table II. High strain fibre, for example, has been developed to offer strength improvements of up to 53% with the corresponding increase in high modulus material being an encouraging 27%.

Against the increasingly stringent composite property requirements being called for, major research effort is being carried out by Hysol Ltd. California (Resin R & D) and Courtaulds Coventry (Fibre R & D) to maintain and improve the Hysol-Grafil range of products. With Hysol-Grafil bringing together the resin expertise of Hysol U.S.A. and the carbon fibre expertise of Courtaulds U.K. the primary research objective is to progress resin and fibre developments together to offer improved composite performance as a synergistic whole rather than as a composite of parts.

This paper deals with some of our carbon fibre research and development directed at improving carbon fibre tensile properties and outlines two of our approaches to achieve this:

1. reducing levels of particulate contamination within the process

   and

2. improving the crystal structure/morphology of carbon fibre filaments.

INTERNAL DEFECTS

The pronounced effect that flaws can have in limiting the ultimate strength of carbon fibres is well documented (1). Both surface and internal flaws have been detected (2, 3) with the latter observed as either foreign particulate contamination, or actual voids within the body of filaments. Detailed research has shown that very often voids result from volatilisation of solid particulate impurities at high temperatures.

On carbon fibre processing the presence of contaminants within the body of the precursor fibre is believed to lead to enhanced crystallisation in the immediate vicinity and resultant misorientation of crystallites. The net effect is a greater likelihood that for a given applied stress the fibre will fail at one of these flaws. Like many others we have carried out detailed inspection of the fracture surfaces of single filaments of carbon fibre after failure in tensile loading. From surface electron micrographs (S.E.M.'s) numerous examples have been observed of tensile failure occurring coincident with flaws.

Against this background and the work reported by Moreton (4) that highlighted the potential property benefits arising from 'clean room' precursor spinning, the last few years have been marked by a major effort by Hysol-Grafil and Courtaulds Research to stringently improve the cleanliness of the 'Grafil' carbon fibre process.

There are essentially three stages to the production of carbon fibre - production of the polymer, spinning of the polymer into a precursor fibre and finally processing of the precursor fibre through to carbon fibre. All of these steps are carried out 'in house' within Hysol-Grafil and are fully researched by Courtaulds Research department. In order to reduce the presence of internal flaws within the final product the R & D effort has been directed at improving filtration in the polymer and precursor stages up to the point of precursor coagulation.

A simplified illustration of the polymerisation process is presented in Figure 1. Like any industrial production process it is impossible to avoid contamination of the final product with small amounts of foreign material. A significant amount of this has its source in pipework and pumps, but a proportion is also unavoidably introduced with some of the chemicals used in the process. To overcome this problem improved filtration has been installed at various points in the process. The solvent feed system now has three stages of filtration going down progressively from 100 to 1μ and below. In addition this is complemented by similar filtration of one of the other ingredients to the reagents make up.

Following this initial work on non-toxic ingredients to the system the work has recently lead to the installation of a particularly stringent filtration stage offering further filtration of the final reagents mix product prior to polymerisation. With the feed product at this stage comprising a mixture of solvent, water, reaction initiator and toxic monomers such as acrylonitrile the installation has had to be designed to rigorous safety tolerances. Despite these restrictions the facility is now installed and removing very substantial quantities of particulate matter from the system.

A simplified diagram of the SAF precursor spinning process is presented in Figure 2. The viscosity of the polymer solution complicates the task of achieving effective filtration. However, complementing the original two primary stages of filtration we have recently installed a final filtration step just prior to filament coagulation.

In the last few years the incremental installation of additional and improved filtration in these two stages of the process has led to the removal of substantial quantities of contaminant material. The overall effect has been to increase the strength of Grafil high strain carbon fibre by 20% clearly illustrating the vital importance of cleanliness in this high technology production process.

CRYSTAL STRUCTURE/MORPHOLOGY OPTIMISATION

Although on a bulk scale the strength of carbon fibre can be severely limited by the presence of inclusions of contaminant material, the presence of a fundamentally homogeneous orientated crystal structure is of equally, or possibly greater importance. This is clearly illustrated by the fact that single graphite whiskers have been produced with a perfectly ordered basal plane crystal structure: in tensile testing products of this type have been found to yield strength and modulus values of 20 and 680 GPa respectively, far above that currently attainable with commercial carbon fibre.

The existence of sheath/core heterogeneity within some carbon fibres has been known for some time (5) with observations confirming the skin to generally have larger, better orientated crystal than the core. In this outer region the layer planes are orientated parallel to the surface. In the core the layer planes are significantly less ordered and can exhibit extensive folding.

With the object of improving the tensile properties of the final carbon fibre product a major programme of work has been carried out by Courtaulds Research to develop a process offering improved homogeneity and orientation of crystallites within carbon fibre filaments. In the last two years, following a technology licencing agreement with F.M.I., this work has progressed from the laboratory bench through to the launch of the new 'Apollo' range of Hysol-Grafil products.

The process involves modification of the steps between SAF spinning and oxidising of the fibre. In initial small-scale laboratory investigations trial batches of fibre were produced offering strength improvements of over 44%. Following this a pilot-production unit was built and work commenced in mid 1984. Commissioning work on this large facility is now complete leading to the development of a family of high strain, intermediate and high modulus 'Apollo' products. In the high strain range material has been produced with strengths above 5 GPa at 255-260 GPa modulus and further development work is in hand to offer improvements to above 5.5 GPa in the near future.

The structure of the new product has been assessed by standard techniques of X-ray diffraction. Results of comparative assessments on standard and Apollo high modulus products are presented in Table III. The azimuthal half-widths, reflecting the degree of orientation of the graphite basal planes parallel to the fibre axis, show significant differences between the two products. In particular the value for the Apollo material is about half that for the standard product, clearly confirming the improved orientation associated with the new product.

In common with the existing range of Hysol-Grafil fibres surface electron microscopy of the Apollo fibre has confirmed the presence of crenellations. These, we believe, are fundamental to achieving good fibre-resin adhesion and optimum property translation through to composite.

In the same way that the new process has given improved carbon fibre tow tensile properties the 'knock on' effect has been to offer corresponding improvement in composite properties. The results of composite strength and modulus testing on standard and Apollo IM samples prepared in Ciba-Geigy resin are presented in Table IV and clearly highlight the better performance of the Apollo grade material.

One of the major objectives of the Hysol Grafil joint venture was to bring together resin and fibre technology to offer the customer optimum performance in composite. With this aim the existing range of products have already been the subject of detailed research to assess the effect of surface treatment level on composite properties (6). In addition, delicate analytical work is being carried out to investigate and fully characterise the physical and chemical nature of the fibre surfaces. In view of this the development work on the Apollo range of products has also involved the specific production of fibre at a variety of surface treatment levels and the detailed analysis of these for surface area, pore size and chemical activity.

Surface area and pore size distribution on the various fibres were determined by nitrogen adsorption techniques following pre-treatment of the samples under vacuum at either room temperature or 130°C. Full details of the analytical techniques have been reported previously by Dr. Robinson of Courtaulds Research (7). The results of analyses on a set of Apollo IM fibres are presented in Table V. In both room temperature and the 130°C series of results the maximum surface area was achieved at the S level of treatment and then tailed off beyond this - presumably due to the rigorous conditions having physically eroded the outer porous layers to reveal more compact inner graphite regions within the fibre structure.

Micropore volume determinations show a similar effect (Table VI) with a maximum level at the S treatment level then falling off at the higher level. In related analysis the lower treated fibres were found to both contain pores in the 36 to 240 Å range. In contrast no pores were detected in the 2S sample.

Finally each of the samples were assessed for chemical activity by determining the extent to which oxygen could be adsorbed on to their surfaces following pre-treatment under vacuum at 1000°C. Interestingly the results presented in Table VII confirm that, not only are surface area and pore size at a maximum at the S level of surface treatment, but so too is chemical activity.

The programme of work on the new process has lead to the development of a new range of 'Apollo' products with improved crystal structure and significant improvements in tensile properties across the full range from high strain to high modulus. In addition, initial work on interphase studies has highlighted the effect of surface treatment level on key fibre physical and chemical properties. This data will form the basis for fully optimising Apollo product process conditions to offer the optimum properties in composite.

REFERENCES

1 Moreton R, Fibre Science and Technology 1 (1968) 273

2 Johnson J W, Applied Polymer Symposia 9 (1969) 229

3 Johnson J W and Thorne D J, Carbon 7 (1969) 659

4 Moreton R and Watt W, Nature 247 (1974) 360

5 Bennett S C and Johnson D J, Carbon 17 (1979) 25

6 Lehmann S, 29th U.S.A. National SAMPE Conference 1985

7 Robinson R, Proceedings of 6th European SAMPE Conference 29 (1985) 145

	1977		1980		1985		(1989)	
Aerospace	88	(34)	265	(36)	1585	(42)	3425	(43)
Sports	130	(51)	320	(44)	1380	(36)	2315	(29)
Others	37	(15)	145	(20)	835	(22)	2250	(28)
Total	255		730		3800		7990	

( ) = % Market Share

Table I World Carbon Fibre Market - Actual and Predicted (Metric Tonnes)

Type	Product Details	Tow Strength GPa	Tow Modulus GPa	% Strain
High Strain	Grafil HT-S (1971)	2.60	230	1.1
	Grafil XAS (1985)	4.00	234	1.7
High Modulus	Grafil HM-S (1970)	2.04	325	0.63
	Grafil HM (1985)	2.60	350	0.74

Table II The Historical Development of Hysol-Grafil Carbon Fibre Properties

	d (002) Spacing Å	Crystal Size L_c Å	Azimuthal Half Width Å
Standard HM	3.39	48.8	23.7
Apollo HM	3.39	60.2	12.1

Table III X-ray Diffraction Data - Apollo HM Fibre

	Composite Strength MPa	Composite Modulus GPa
Grafil HS[1)]	1800	135
Apollo IM[2)]	2320-2450	166-168

1) Ciba-Geigy 914 resin

2) Ciba-Geigy 6376 resin

60% Vol Fraction

Table IV Composite Test Results

Sample	Surface Treatment Level	BET Surface Area Nitrogen Adsorption m^2/gm	
		R.T. Evaluation	130°C Evaluation
Apollo A	Zero	0.750	1.387
Apollo B	S	1.168	1.519
Apollo C	2S	0.708	0.870

Table V Surface Area Analysis - Apollo IM Fibre

Fibre Sample	Surface Treatment Level	Micropore Volume $\mu g\ N_2/gm$ Fibre	Mesopore Size Range
Apollo A	Zero	100	36 to 240 Å
Apollo B	S	110	36 to 240 Å
Apollo C	2S	45	None Detected

Table VI Pore Volume and Size Analyses - Apollo IM Fibre

Fibre Sample	Surface Treatment Level	Oxygen Complex Formed μ moles/gm	Distance Apart of Active Sites Å
Apollo A	Zero	0.33	26.3
Apollo B	S	2.0	11.2
Apollo C	2S	0.75	13.8

Table VII Active Area Analysis - Apollo IM Fibre.

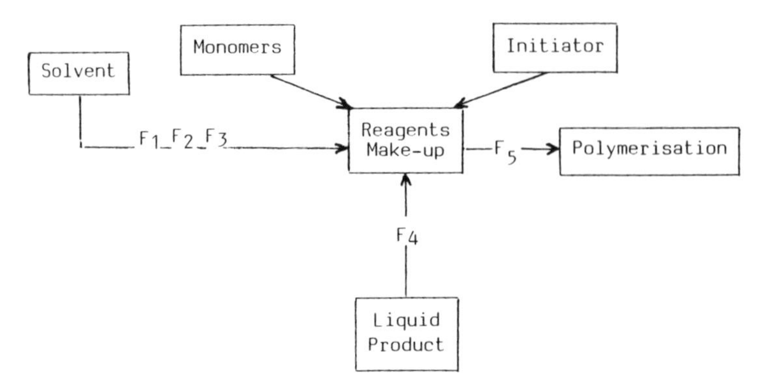

F - Filtration Stages

Figure 1 Polymerisation Process

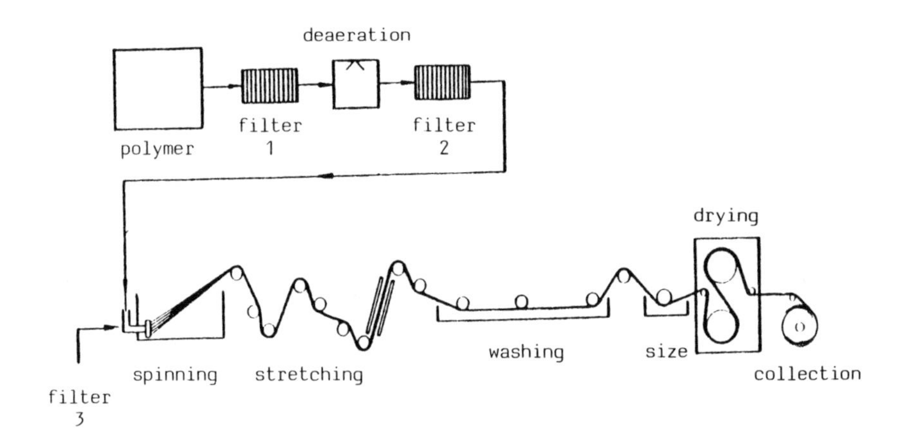

Figure 2 SAF Precursor Spinning

A COMPARISON OF AVAILABLE CARBON FIBERS

Donald R. Lovell

Materials Consultant
HiPerTech, 'Reapers Croft'
Chilcompton, BATH BA3 4HB

A list is given of 17 manufacturers of High Modulus Carbon Fibre (US - graphite fiber) worldwide who make 74 grades. These grades are categorised in modulus and strength groups, UHM,HM,IM,VHS,HS; classified by number of filaments/tow and weight/unit length, and ranked according to their specific strength, specific modulus and strain to failure.

It is suggested that the preparation of international standards for groups of grades of similar modulus would be an advantage.

INTRODUCTION.

The history of carbon fibre manufacture for the last twenty years has been a story of constant change in the products. This has been called 'improvement' by the suppliers although some designers and users would not agree that the introduction of new grades at frequent intervals was an advantage.

Initially made from rayon by hot stretching during graphitisation, a commercial product (Thornel 50) achieved strength and modulus values of 2.76 GPa and 345 GPa twenty years ago, which is higher strength than is achieved at this modulus today (because the strength of rayon-based fibre increases with modulus), but the yield was only about 15% w/w of the precursor rayon.

The use of polyacrylonitrile (PAN) as a precursor allowed the stretching to be carried out at low temperature and increased the yield to 50%.

The first PAN-based grades were heated at 1200 - 1400°C for High Strength or 2200 - 2500°C for High Modulus fibres and gave, respectively, strength/modulus values of 2.3/230 and 1.8/415 GPa. Later, a cheaper High Strength grade was introduced (2.1/190) and an oxidation-resistant grade (2.2/280) for use with polyimide resins.

The main changes which have occurred since the early 1970's have been

a) reduced coefficient of variation of strength and modulus for each grade

b) an increase in strength at the same modulus

This has lead to the Very High Strength (VHS) or High Strain grades which achieve strength/modulus values around 4.3/230 GPa and thus a strain close to 2%; and the Intermediate Modulus (IM) range with strength/modulus values around 4/285 GPa made with a reduced diameter of 4 - 5 micrometres by hot stretching. This modulus can be obtained by higher temperature processing of a standard 7 - 8 micrometre diameter fibre, as was done for the oxidation-resistant grade in the early 70's, but the high strength cannot be obtained by this process so the smaller diameter fibre has to be produced.

c) ultra-high modulus (UHM) PAN-based grades

These have modulus values greater than 395 GPa. Values of 586, 517, 441 or 400 GPa are available with strengths around 1.8 GPa from the four suppliers who make them.

d) the development of the process using pitch as a precursor.

Pitch-based fibres have been developed in two classes of carbon fibre:

i) High Modulus fibres to compete with PAN-based grades

ii) Low Modulus (Fibrous Carbon & Graphite) fibres for use as furnace insulation, pump packings and as filler for ablative phenolic resin mouldings.

Pitch precursor provides a higher yield than PAN, up to about 80%, and the process can produce fibres with the highest modulus of any carbon fibre - up to 826 GPa, approaching the 1000 GPa of a graphite crystallite. The strength is about 2.2 GPa, which makes these fibres most suitable for stiffness-critical structures.

The Low Modulus fibres are much cheaper and they typically have strength/modulus values of 0.7/35 GPa with all the electrical and thermal properties of carbon or graphite in a fibrous form.

TABULATED DATA.

The tables list

TABLE I. Manufacturers & Trade Names
II. Grades & Categories
III. Classification by Filaments/Tow & Weight/unit length
IV. Ranking by Specific Modulus
V. Ranking by Specific Strength
VI. Ranking by Strain

In Tables II and III the grades are categorised into five classes defined as:

Ultra-High Modulus (UHM) - Modulus over 395 GPa
High Modulus (HM) - Modulus 310 - 395 GPa
Intermediate Modulus (IM) - Modulus 255 - 310 GPa
Very High Strength (VHS) - Strength over 3.5 GPa & Modulus under 255 GPa
High Strength (HS) - Strength up to 3.5 GPa

No attempt has been made to rank fibres by actual strength since this property is changing rapidly with increases being announced for each grade at frequent intervals. The ranking by Specific Strength is likely to become out-of-date quickly for this reason but is included to indicate the range of values current.

DISCUSSION.

The very large number of grades, spanning a 3/1 range of specific modulus and a 3½/1 range of specific strength, can be rationalised by grouping and it is suggested that it would be helpful to both users and manufacturers for international standards to be agreed for groups of grades. For example, the modulus values can be grouped as shown in Table VII to provide alternative sources of fibres with moduli which are within ± 2½% of the mean for the widely-used 223 - 247 GPa range and ±5% for the remainder, excluding the 'special' UHM grades and the two with moduli under 200 GPa.

This reduces the bulk of the grades to 7 groups with modulus values in each group which are sufficiently close, considering the coefficient of variation of this property, to be suitable for design purposes.

CONCLUSIONS.

The variation in strength and modulus between grades is large, justifying categorising them into five classes, but it is suggested that grouping grades by modulus and agreeing international standards for each group could assist designers and users as well as manufacturers.

Donald R Lovell graduated in Chemical Engineering from Loughborough College. He worked for the A.P.V. Co. Ltd and the South Eastern Gas Board before joining the Morgan Crucible Group in 1957. After some years in charge of the Furnace Design section in Morganite Research & Development Ltd he became Manager of the Process Engineering Department and then the New Products Department. When the manufacture of High Modulus Carbon Fibre was started he was successively Development, Production, Quality Control and Applications Manager in Morganite Modmor Ltd. Since 1978 he has been an independent Materials Consultant on high performance fibres and composites trading under the name HiPerTech. He compiled the 'World-wide Carbon & High Performance Fibres Directory now in its 3rd Edition, and contributes to and distributes for Composite Market Reports Inc, the Key Personnel Lists for USA and Europe for the Advanced Composite Materials industry.

****

COMPANY	COUNTRY	TRADE NAME
Afikim Carbon Fibers	Israel	ACIF
Avco Specialty Materials Divn.	USA	Avcarb
Celanese Corporation	USA	Celion
Enka AG	Germany	Tenax
Great Lakes Carbon Corporation	USA	Fortafil
Hercules Incorporated	USA	Magnamite
Hitco Materials Divn,Armco Inc.	USA	Hi-Tex
Hysol Grafil Ltd	GB	Grafil
Mitsubishi Rayon Co Ltd	Japan	Pyrofil
Nippon Carbon Co Ltd	Japan	Carbolon
R.K.Carbon Fibres Ltd	GB	RK Carbon
Sigri Elektrographit GmbH	Germany	Sigrafil
Soc. Fibres de Carbone(SOFiCAR)	France	Torayca
Stackpole Fibers Co. Inc.	USA	Panex
Toho Beslon Co Ltd	Japan	Besfight
Toray Industries Inc.	Japan	Torayca
Union Carbide Corporation	USA	Thornel

TABLE I. MANUFACTURERS OF CARBON FIBRE TOW.

TRADE NAME.	UHM	HM	IM	VHS	HS
ACIF	-	HM	-	-	HT,XHT,IS
Avcarb	-	-	-	-	G-160
Celion	GY 80	G-50	G-40	3000ST 6000ST 12000ST	1000 3000 6000 12000
Fortafil	-	F-5(o)			F-3(o) & F-3(c)
Magnamite	-	HMS, HMU	IM6	AS6	AS4,AS1
Hi-Tex	-	-	-	HS grades	3000,6000,10000, 12000
Grafil	-	HMS, HM*	IMS,IM*	XAS(HS),HS*	XAS, XAS(HP),A-3
Pyrofil	-	M-1	M-2	T-2	T-1
Carbolon	-	-	-	-	Z-3
RK Carbon	-	HiM	-	RK35	RK30, RK25
Sigrafil	UHM	HM	-	-	HF, NF
Panex	-	-	-	-	30,30R, 30Y
Besfight/Tenax /Tenax J **	HM45	HM40,HM35	HM28,IM- 400,IM- 500	ST 1,2,3	HTA
Torayca	M-46	M-40	M-30 T-800	T-700 T-400	T-300
Thornel	P 120 P 100 P 75S	P 55 P 55S T-50	T-40	T-700 T-600 T-500	T-300, P 25

* Grafil 'Apollo' grades

** Tenax & Tenax J from Enka AG; Besfight from Toho Beslon Co Ltd

Thornel P grades are pitch-based

See Table I for Company names

TABLE II. GRADES AND CATEGORIES OF CARBON FIBRES.

FILAMENTS PER TOW	TEX (g/km)	YIELD (m/kg)	GRADE
Ultra-High Modulus			
384	42	23809	Celion GY80SE Celion GY70SE
1000	160	6250	Thornel P 55
2000	317	3154	Thornel P 120 Thornel P 100
	320	3125	Thornel P 75
3000	180	5555	Sigrafil UHM
6000	360	2777	Besfight HM45 (Tenax HM45)
12000	740	1351	Besfight HM45 (Tenax HM45)
	750	1333	Sigrafil UHM
High Modulus			
1000	61	16393	Torayca M 40
	75	13333	Magnamite HMU
2000	320	3125	Thornel P55S
3000	180	5555	Besfight HM35 (Tenax HM35) Sigrafil HM
	182	5494	Torayca M 40 Thornel T 40
	190	5263	Acif HM
	226	4427	Magnamite HMU
4000	714	1400	Thornel P55S
6000	364	2747	Thornel T50 Torayca M 40
	370	2703	Besfight HM 40 Besfight HM 35 (Tenax HM 40,35)
	380	2631	Acif HM RK Carbon HiM Grafil HMS
	454	2202	Magnamite HMU
10 000	826	1211	Magnamite HMS
	900	1111	Pyrofil M-1 Grafil HMS
12 000	750	1333	Besfight HM 40 Besfight HM 35 (Tenax HM 40,35) Sigrafil HM
	760	1316	Acif HM
	770	1299	Celion G 50
High Modulus (continued)			
12 000	909	1100	Magnamite HMU
40 000	3502	285	Fortafil F-5
160 000	14008	71	Fortafil F-5
Intermediate Modulus			
3000	200	5000	Pyrofil M-2 Celion G 40
6000	322	3105	Thornel T 40
	375	2606	Grafil IMS
	400	2500	Pyrofil M-2 Celion G 40
12 000	760	1316	Besfight HM28 (Tenax HM28)
	765	1307	Grafil IMS
	800	1250	Pyrofil M-2
	817	1224	Celion G 40
	?	?	MagnamiteIM6
Very High Strength			
3000	198	5050	Torayca T300 Torayca T400 Thornel T500
	200	5000	Pyrofil T-2 Besfight ST3 (Tenax ST3)
	201	4975	Celion 3k ST
6000	315	3175	Thornel T700
	390	2564	RK Carbon 35
	396	2525	Torayca T400 Thornel T500
	400	2500	Grafil XAS(HS) Pyrofil T-2
	404	2475	Celion 6kST
	410	2439	Besfight ST3 (Tenax ST3)
12 000	780	1282	RK Carbon 35
	794	1259	ThornelT500
	800	1250	Pyrofil T-2 Grafil XAS(HS) Torayca T400 Torayca T300
	810	1234	Besfight ST3 (Tenax ST3)
	817	1224	Celion 12kST
	?	?	MagnamiteAS 6

TABLE III. CLASSIFICATION OF CARBON FIBRES.

FILAMENTS PER TOW	TEX (g/km)	YIELD (m/kg)	GRADE
High Strength			
1000	66	15151	Sigrafil HF&NF
			Besfight HTA
			(Tenax HTA)
			Thornel T300
	67	14925	Celion 1000
3000	195	5128	Acif IS,HT,XHT
			RK Carbon 25,35
	198	5050	Thornel T 300
	200	5000	Carbolon Z3
			Pyrofil T-1
			Sigrafil HF,NF
			Besfight HTA
			(Tenax HTA)
	201	4975	Celion 3000
	211	4739	Hi-Tex 3000
	220	4545	Magnamite AS4
4000	622	1610	Thornel P25W
6000	390	2564	RK Carbon 25,30
	391	2557	Acif IS,HT,XHT
	396	2525	Thornel T 300
	400	2500	Grafil XAS
			Grafil XAS(HP)
			Pyrofil T-1
			Sigrafil HF,NF
	404	2475	Celion 6000
	410	2439	Besfight HTA
			(Tenax HTA)
	433	2309	Hi-Tex 6000
	440	2273	Magnamite AS 4
	450	2202	Carbolon Z3
10 000	711	1406	Hi-Tex 10000
	750	1333	Carbolon Z3
	806	1241	Magnamite AS 1
	950	1053	Grafil A-3
12 000	780	1282	RK Carbon 25,30
	781	1280	Acif IS,HT,XHT
	794	1259	Thornel T 300
	800	1250	Pyrofil T-1
			Grafil XAS(HP)
			Grafil XAS
	810	1234	Sigrafil HF,NF
			Besfight HTA
			(Tenax HTA)
	881	1135	Magnamite AS 4
			Celion 12000
	900	1100	Hi-Tex 12000
High Strength (continued)			
24 000	1610	621	SigrafilHF, NF
	1630	603	Besfight HTA
			(Tenax HTA)
40 000	2600	385	RK Carbon 25
	3702	270	Fortafil F-3
	?	?	Acif IS
160 000	14444	69	Avcarb G-160
	14880	67	Fortafil F-3
	14889	67	Panex 30
320 000	20800	48	RK Carbon 25
	29778	33	Panex 30
	?	?	Acif IS

NOTE: Data for Magnamite IM6 & AS 6 not available

TABLE III. CLASSIFICATION OF CARBON FIBRES. (continued)

SPECIFIC MODULUS (GPa)	TRADE NAME & GRADE	SPECIFIC MODULUS (GPa)	TRADE NAME & GRADE
Ultra-High Modulus		High Modulus	
379	Thornel P 120	220	Grafil HMS(6k)
337	Thornel P 100	216	Torayca M 40
299	Celion GY 80	215	Thornel T 50
264	Celion GY 70	214	Besfight HM 40
260	Thornel P 75		(Tenax HM 40)
256	Torayca M 50	208	Magnamite HMU
232	Besfight HM 45	201	Celion G 50
	(Tenax HM 45)	193	Besfight HM 35
200	Sigrafil UHM		(Tenax HM 35)
		191	Pyrofil M-1
		190	Thornel P 55S
Intermediate Modulus		189	Magnamite HMS
		183	RK Carbon HiM
171	Thornel T 40	183	Grafil HMS(10k)
165	Magnamite IM 6	180	Acif HM
163	Torayca T 800H	173	Torayca M 30
162	Celion G 40	170	Sigrafil HM
160	Besfight HM 28		
	(Tenax HM 28)		
142	Pyrofil M-2		

TABLE IV. RANKING BY SPECIFIC MODULUS.

SPECIFIC STRENGTH (GPa)	TRADE NAME & GRADE	SPECIFIC STRENGTH (GPa)	TRADE NAME & GRADE
Very High Strength		High Strength	
2.5	Thornel T 700	2.1	Magnamite AS 4
2.4	Celion ST grades	2	Celion(1,3,6,12k)
	Besfight ST3	1.95	Pyrofil T-1
	(Tenax ST 3)	1.9	Grafil XAS(HP)
2.3	Thornel T 600	1.85	Acif XHT
	Besfight ST 2	1.81	Besfight HTA
	(Tenax ST 2)		(Tenax HTA)
	Pyrofil T-2		Carbolon Z3
	Thornel T 400	1.78	Hi-Tex (3,6k)
	Magnamite AS 6	1.72	Grafil XAS
2.15	Grafil XAS(HS)		Hi-Tex(10,12k)
2	Thornel T 500		Magnamite AS 1
	Besfight ST 1	1.66	Avcarb G-160
	(Tenax ST 1)	1.62	Acif HT
	Torayca T 300		Fortafil F-3
	Hi-Tex HS grades	1.44	Panex 30
		1.4	Acif IS
		1.28	Grafil A-3
		0.74	Thornel P 25W

NOTE: RK Carbon grades & Sigrafil HF & NF not ranked. (Minimum values of strength only provided)

The Specific Modulus of all VHS grades is 133 ± 4 GPa.

The Specific Modulus of all HS grades is in the range 125 - 132 GPa

TABLE V. RANKING BY SPECIFIC STRENGTH.

% STRAIN	TRADE NAME & GRADE
2	Grafil Apollo HS
1.84	Celion ST grades
1.83	Thornel T 700: Besfight (Tenax) ST3
1.80	Torayca T 400H; T 800H
1.75	Besfight(Tenax) ST 2
1.72	Thornel T 600
1.70	Celion G-40
1.67	Pyrofil T-2
1.65	Magnamite AS-6
1.64	Grafil XZS(HS)
1.60	Tenax IM 500X
1.55	Acif XHT
1.54	Besfight(Tenax) ST 1 : Hi-Tex HS grades
1.52	RK Carbon 35
1.51	Thornel T 500

% Strain = Strength/modulus x 100 (GPa)

TABLE VI. RANKING OF VHS FIBRES BY STRAIN.(%)

GROUP Mean Modulus & Limits	FIBRE MODULUS (GPa)	TRADE NAME & GRADE
Group HM 1	392	Torayca M-40: Besfight(Tenax) HM 40
372.5 ± 5%	390	Thornel T 50: Grafil Apollo HM
(354 - 391)	380	Thornel P 55S
	379	Magnamite HMU
	370	Grafil HMS(6k)
	358	Celion G-50: Besfight(Tenax) HM 35
	353	Pyrofil M-1
Group HM 2	345	Mangamite HMS: Besfight(Tenax) HM 35
337.5 ± 2½%	340	Grafil HM(10k):Sigrafil HM:Fortafil F-5
(329-346)	335	Acif HM
	330-350	RK Carbon HiM
Group IM 1	306	Pyrofil M-2
290.5±5%	297	Thornel T 40
(276-305)	295	Tenax IM 400, IM 500
	294	Torayca T 800H, M-30
	290	Grafil IMS
	280	Grafil Apollo IM
	275	Celion G-40
Group IM 2	274	Besfight(Tenax) HM 28
261±5%	262	Panex 30R, 30Y
(248-274)	250	Torayca T 400H
	248	Thornel T 700
Group VHS 1	245	Pyrofil T-1: Grafil Apollo HS
240 ±2½%	241	Thornel T 600
(234 -246)	240	Thornel T 500: Carbolon Z3-2000: Sigrafil HF
	238	Grafil XAS; Besfight (Tenax) HTA,ST3: Thornel T 400: Torayca T-300: Magnamite AS-4 Pyrofil T-2 ((RK Carbon 25,35))*
Group VHS 2	234	Celion(1,3,6,12k)
230.5 ±2½%	230	Torayca T-300: Carbolon Z3-4500/7500/9000: Acif IS,HT,XHT: ((RK Carbon 25,35))*
(225 - 236)	228	Torayca T-300: Magnamite AS-1
	227	Fortafil F-3: Hi-Tex(3,6,10,12k)
Group HS1	221	Avcarb G-160:
210.5 ± 5%	220	Sigrafil NF: Panex 30: ((RK Carbon 25,35))*
(200 - 221)	200	Grafil A-3

NOTE:*RK Carbon 25 & 35 - modulus quoted as 215 - 240 GPa

TABLE VII. PROPOSED MODULUS GROUPS.

PROPERTY TESTING AND ASSESSMENT OF CFRP

G. Dorey

Royal Aircraft Establishment

Paper not available at press time.

Graham Dorey graduated from Trinity College Cambridge with an Honours Degree in Natural Sciences (Physics) and joined RAE where he started work on hardening mechanisms in copper. This led to the development of dispersion hardening in aluminium, in collaboration with the University of Surrey, for which he obtained his PhD. Carbon Fibres, developed at RAE, were then becoming of increasing interest and he studied the problem of impact damage in CFRP. For the past 9 years he has led the Composite Materials Section covering static and fatigue properties, fracture characteristics, effects of impact damage and environment, the development of improved fibres and detailed investigations of interactions between fibres and matrix. He has sat on a number of international committees concerned with damage tolerance and airworthiness of composite structures. He is an Honorary Visiting Reader in the Department of Metallurgy and Materials Technology at the University of Surrey.

Part 2

Intermediate Materials Conversion and Processing

WOVEN CFRP FABRICS

P. Lissak

Stevens-Genin

TAPE or FABRIC ?

What kind of carbon fiber reinforcement to select for a laminate ?

About 10 years ago, the above question was often asked by the engineering teams, after both TAPE and FABRIC had been successfully used in composite parts. It seemed by then that only one of them would survive.

What is the present state-of-the-art ?

STEVENS-GENIN - a long established Firm whose name is HEXCEL-GENIN as of 1st October 1985 - pioneered the weaving of carbon fibers as early as 1968. The WEFTAMATIC technique was then used to weave the carbon yarns introduced by UNION CARBIDE and to develop weftwise UniDirectional fabrics, width 107cm, weight 150 to 300 g/m2. One will remember that a WEFTAMATIC fabric was used by SNIAS to manufacture the blades of their "Dauphin" SA360 helicopter, an helicopter which beat the world speed record in its category in 1971. But the WEFTAMATIC technique allowed the use of fine and costly threads only, and had an unchangeable 107cm width which generated waste when the fabric had to be cut out and laid up at 45° angle. Both conditions rendered WEFTAMATIC fabrics non competitive vs. prepregged UniDirectional tapes made from cheaper 10 000 filament carbon yarns.

UD tapes thus developed from 1972 onwards, because the only "reasonable" priced yarns were the SEROFIM and the COURTAULDS 10 000 filament-yarns, which unfortunately could not be woven into conventional fabrics.

In 1972 the 1000 and the 3000 filament high resistance TORAYCA yarns were introduced in Europe, and the weaving trials started again.

In 1974 STEVENS-GENIN had a consistent range of carbon fabrics that met the needs of their clients.

Today, most of the carbon fabrics are woven from 3000 filament yarns whose weaving is well mastered. For economical reasons a trend towards the use of 6000 filament yarns is to be noticed, while 10 000 filament yarns are in lesser demand. Selecting from US TAPES and WOVEN FABRICS is no easy task for the laminator who is seldom aware of the possibilities offered by each of these techniques.

1. UD TAPES

On the present market, two types of tapes are to be found DRY TAPES and PREPREGGED TAPES.

The widthwise cohesion of the DRY TAPES is ensured by a thermoplastic binder. Such tapes are used for making laminates by wet moulding, or contact moulding, a technique that is not convenient for production of high performance laminates.

The widthwise cohesion of the PREPREGGED TAPES is ensured by the laminate's resin. Such tapes are used for making high performance laminates, prepregs being specially suitable for autoclave and hot press moulding.

All types of carbon can be used : HM, HT, IM.

The standard available widths vary from 300 to 600mm, with a trend towards larger widths (1000mm). The resin contents are between 35 and 45% (in weight).

The available weights - expressed in g of dry carbon per m2 - are ranging from 60 to 350.

1000 filament yarns are used in the lighter weights (60 to 120 g/m2), 3000 filament yarns in the medium range of weights (100 to 200 g/m2), 6000 filament yarns in the heavier weights (140 to 350 g/m2).

In each category, no twist yarns are used for the lighter weights, and twisted yarns for the heavier weights.

2. CONVENTIONAL FABRICS

We define a conventional fabric as : a cloth produced by interlacing two sets of yarns whose elements pass each other essentially at right angles, one set of yarns being parallel to the fabric axis.

For a closer definition of a fabric, one must state :

- the nature of the yarns used lengthwise and weftwise : in the case of carbon, the type of carbon and the number of filaments in one yarn.
- the average count of warp ends and filling picks.
- the weave, i.e. the pre-defined pattern according which the warp and filling yarns are interlaced (Plain, Satin, etc.).

The weight of the fabric results of the above. The weight tolerance depends directly upon the weight tolerance for the yarns.

(a) Mechanical characteristics of a fabric

Determination of mechanical characteristics on a dry fabric do not provide any meaning element. This determination must be done in a laminate.

The two parameters that have an effect upon the mechanical performance are

- the weave : an 8-H-Satin will perform better than a Plain fabric of same weight made from similar yarns. The fabric count, and the yarn count, are also important factors.
- the resin content (as in all laminates). But a 200 g/m2 Plain fabric made from 3000 filament yarns will require a higher resin content than a 400 g/m2 8-H-Satin.

(b) Physical characteristics of a fabric

Beside the thickness and the aerial weight of the fabric, the following 4 parameters may help define the properties of a carbon fiber fabric meant for technical end-uses :

- the cover-factor, closely related to the fabric appearance, and defined as the area covered by the yarns inside the fabric. The value 1 is to be aimed at, or the fabric would have the appearance of a grid.
- the porosity, depending on the fabric thickness, and allowing to calculate the free space inside the fabric, a volume that will be filled by the resin during the laminating operation. To achieve good mechanical properties, the porosity must be lower that 50% in volume.
- the permeability to resins, which depends on the weave and on the wetting-out capacity of the fiber (its sizing). The highest possible permeability is to be aimed at.
- the drape, defining the conformability of the fabric. The 12-Satin and the 8-Satin are the best conformable weaves. But in the case when the laminate to be achieved is a plane one, this "drape" notion can be overlooked.

(c) Range of weights for carbon fabrics

First, one must tell apart BALANCED FABRICS, and UD FABRICS.

The range of BALANCED fabrics spreads from 100 to 750 g/m2. Here again, as in the US Tapes, 1K or 3K or 6K yarns will be used.

The range of UD fabrics is more limited : from 100 to 320 g/m2.

(d) Hybrid Fabrics

Hybridizing consists in mixing yarns of different nature, for instance Carbon and Glass. Hybridizing is possible in woven fabrics only. Weaving lends itself to such operation as far as the yarns in question are of similar volume, though of different nature.

From the mechanical point of view, the modulus of hybrid fabrics will follow the law of the mixtures according to the volumic percentage of the components. But when it comes to tensile strength, they exhibit lower values than their components, because their breaking elongations are different and one material will break before the other one.

3. COMPARISON BETWEEN TAPE AND FABRIC

(a) Mechanical performance

One could look at a balanced fabric as being equivalent to 2 plies of tapes laid at 90° angle. In fact, laboratory tests on specimens show that such an equivalence is not reached and that the modulus of a fabric is 5% lower, and its breaking resistance 10% lower.

In large manufactured composite parts, the gap between fabrics and tapes is less sizeable, because the lower but evenly distributed characteristics of fabrics are offset by the local defects featured by tapes.

The fiber content is 3% higher in tapes, on an average.

For a given resin, the resistance to fatigue is higher in tapes. Whereas fabrics behave better towards impact or stress.

(b) Handling

The dimensional optimization is easier when using tapes. Anyhow, such an optimization is possible with fabrics, if both balanced and UD fabrics are used in the same laminate.

Tapes need only to be placed side by side so that the stress in the part be uninterrupted. Fabrics require an overlapping to get crosswise (weftwise) continuity.

Hand-draping of contoured parts is easier with fabrics because of their more reliable consistency and their shorter draping lead-time. But machine-draping is easier with tapes.

Automatic cutting-out (used for preparing kits) will prove more interesting with fabrics whose higher width will generate less waste.
And last, after the curing is completed, a laminate made from fabric will lend itself better to sanding and machining operations.

4. CONCLUSIONS

Producers and end-users of carbon composite parts believe that both products have found their niche in different applications.
Therefore, we no longer ask : Tape OR Fabric ?
We say : Tape AND Fabric.

PAPERS—A NEW DIMENSION IN CARBON FIBER MATERIALS

N.J. Walker

Fibertec

Carbon fibre in the form of paper is a relatively new type of material. Papers offer the opportunity to produce a sheet material which is lightweight, inexpensive and has radically different properties to the more usual woven and unidirectional fabrics.

Some applications of carbon fibre paper relating to electrical thermal and chemical properties are briefly discussed. Structural properties of these materials are virtually unknown. Some ideas are presented regarding possible structural uses.

INTRODUCTION

Paper is produced by slurrying fibres in a liquid and allowing the slurry to drain through a continuously moving mesh. The resulting web is removed from the mesh and any remaining liquid removed by pressing and heating (see fig 1). Papers can range from veils or tissues of around 20 grammes per square meter (g/m^2) to thick felts in excess of 250 g/m^2.

The common characteristic of papers is that they consist of a two dimensional sheet of short fibres with a totally random arrangement within the plane of the sheet (fig 2).

Since its invention in China around 105 AD paper has been produced from natural cellulose fibres; these fibres have the ability to form strong interfibre bonds by hydrogen bonding following mechanical treatment in water. Since carbon fibre has no natural bonding ability, it is necessary to bond the fibres by the use of a chemical binder. A wide range of thermoplastic and thermosetting binders are available. Binders are chosen for compatability with final resin systems and to give desired physical properties such as stiffness and strength of the sheet. In all cases binder content represents less than 10% of the total weight (see fig 2). To achieve a good dispersion of fibre in liquid a short chopped fibre must be used; the maximum fibre length used at present is 25 mm.

Hybrid papers can be produced very easily and accurately by mixing fibres in the wet slurry stage. Most fibres which are available in suitable lengths can be used to produce hybrid papers. The most common hybrids at present are carbon/glass, carbon/aramid and carbon/cellulose. Papers produced in this manner are true hybrids in that they consist of two inter-penetrating random fibre networks.

PROCESSING

Carbon fibre papers behave in a similar way to glass surfacing tissues or at heavier weights to chopped strand mat. They can be cut to shape with scissors and are strong enough to be continuously unwound from a roll in processes such as SMC production.

At lighter weights (below 100 g/m^2) papers are highly deformable and can fit complex mouldings more readily than woven products. Thermoset composites using carbon fibre papers can be produced using any of the methods which allow pre-placement of the reinforcement in sheet form and in continuous processes such as pre-preg production and pultrusion. Thermoplastic composites can be produced by film stacking techniques.

USES OF CARBON FIBRE PAPERS

The uses of carbon fibre papers can be divided according to the property of the fibre which is being exploited:

Electrical conductivity	- antistatic products - EMI/RFI shielding - resistive heating
Thermal conductivity	- inductive heating
Chemical resistance	- improved FRP chemical plant
Strength and Modulus	- structural applications

ANTISTATIC FRPs USING CARBON FIBRE PAPERS

The increasing use of sophisticated electronic equipment coupled with the greater use of plastic materials has recently highlighted the problems of static electricity. Static is a particular hazard in industries where volatile gases or finely divided dust is present because of the danger of static induced explosion. Static is a severe problem in the microelectronics industry since expensive components can be irreparably damaged by static charges as low as 50V.

The problems of static electricity can be largely overcome by ensuring that equipment in static sensitive environments has sufficient surface conductivity to bleed away any static charges before they can become dangerous.

The effectiveness of antistatic or, more properly, static dissipative materials is indicated by the measurement of surface resistivity. Surface resistivity is measured in ohms per square ($\Omega\,\square$) and is the resistance measurement between electrodes which form two sides of a square (see fig 3). An alternative measurement is static decay time - this is the time taken for a static voltage (usually 5000V) to decay to 10% of its original value. Static decay time is usually directly proportional to surface resistivity.

Existing specifications for static protection materials usually require a surface resistivity of less than $10^5\Omega\,\square$ for a material to be considered conductive (refs 1,2). To achieve the desired levels of surface resistivity it is seldom necessary to use a pure carbon fibre product; hybrids are used to dilute the carbon fibre with a cheap non-conductive fibre. The level of carbon fibre required to give a specified resistivity is dependent on the nature of the non-conductive fibre and on the aspect ratio of the carbon fibre [Bigg and Stutz (3)]

A long standing relationship between papermakers and plastics producers has existed in the production of decorative laminates for furniture or building panels using melamine formaldehyde resins. Cellulosic papers containing carbon fibre are used to produce a laminate with a surface resistivity of 10^2 to $10^4 \Omega\square$ for the production of static safe furniture, particularly work benches for electronics assembly. The low level of carbon fibre needed to achieve this resistivity is illustrated in fig 4.

A hybrid paper of "E" glass and carbon fibre is used to provide a conductive surface to GRP plant for use in materials handling particularly for chemical processing and the handling of dry powders. GRP ducting and vessels can be produced with a conductive surface on the inside or outside or both depending on where static protection is needed. When producing static dissipative GRP by this method, thick gel coats should be avoided; this is because a small number of fibres need to penetrate the resin surface in order to provide contact points between the surface and the conductive layer. To maintain electrical continuity individual sections of antistatic GRP should be wired across joints and the whole assembly should be securely earthed.

EMI/RFI SHIELDING

A major problem in the use of platics as a housing for electronic components is the one of its transparency to unwanted electrical signals in the forms of electromagnetic or radio frequency interference (EMI/RFI).

Effective shielding against EMI is achieved by housing electrical equipment in a conductive enclosure. Traditional sheet metal housings provide excellent shielding but modern plastic enclosures require difficult and often expensive after treatments such as direct metallisation or the application of conductive coatings. The provision of EMI shielded enclosures for computers is now required by law in the USA and similar legislation for Europe is expected shortly. The requirements laid down by the Federal Communications Commission (FCC) are given below:

TABLE 1 - F.C.C. Rules and Regulations Part 15

F.C.C. Class	Definition	Frequency MHz	Emission Limits Conducted $\mu V m^{-1}$	Emission Limits Radiated*
A	Devices for use in commercial business or industrial environments	0.45 - 1.6	1000	
		1.6 - 30	3000	
		88 - 215		50
		216 - 1000		70
B	Portable equipment for use in residential environments	0.45 - 30	250	
		30 - 88		100
		88 - 216		150
		216 - 1000		200

* Class A measured at 30m, Class B at 3m

Shielding effectiveness or attenuation is measured in decibels and is the logarithmic ratio of the energy which penetrates the shield to the energy of the incident wave. A typical shielding level of 35 dB means that in excess of 99.9% of the wave energy is stopped. For most applications a shielding level of 30 to 40 dB is needed. Lower levels may be acceptable when only electromagnetic compatability (EMC) is required. Higher levels are needed for highly sensitive equipment.

The level of shielding effectiveness for sheet materials may be predicted from the following equation [Reilly (5)]:

$$SE\ dB = 20 \log \frac{Zw}{4Zb}$$

Where SE = attenuation in decibels

Zw = impedence of air - approx 377

Zb = impedence of shield in ohms per square (surface resistivity)

The actual attenuation achieved is also dependent on the frequency of the radiation and the thickness of the shield. However, the above equation usually gives a reasonable approximation. The surface resistivity of a typical high strength, PAN based carbon fibre paper is $2\,\Omega\square$ which when substituted into the above equation gives a predicted level of attenuation of 33 dB. Higher levels may be obtained by using high modulus or metal coated fibres.

In summary, carbon fibre papers offer an economical means of providing EMI shielding in certain types of plastics. Since the papers form an integral part of the structure shielding cannot be lost by abrasion, accidental damage or delamination. The presence of a carbon fibre does not adversely affect the physical properties of the composite and may, in fact, give some improvement.

RESISTIVE HEATING

By utilising both the electrical and thermal characteristics of carbon fibre, a heating layer can be introduced to reinforced plastic components. The principle used is that of resistive heating in which a current passing through a resistor causes an increase in temperature. Heating panels are produced by laminating a carbon fibre layer between two layers of insulator (glass) in a suitable resin. A demonstration plate has been produced as follows:

900 g/m^2 CSM/ 60 g/m^2 carbon fibre paper/ 900 g/m^2 CSM in Derakane 470-36 vinyl ester resin

Brass electrodes were connected to either end of the carbon fibre layer, overall dimension of the panel were:

length 250 mm, width 57 mm, thickness 5 mm, separation of electrodes 200 mm, resistance between electrodes 10 (see fig 5).

The relationships between time and temperature at 12V and between applied voltage and maximum temperature at 1A were determined and are given in figs 6 and 7.

It will be seen from figure 6 that the temperature rises gradually, stabilises and remains steady. The maximum temperatures which can be obtained are likely to be dependent on the heat resistance of the resin but are expected to be in excess of 150°C. The thermal conductivity of the carbon fibre together with the isotropic nature of the paper ensures that heating is even throughout the component. Among the possible applications for these products are:

Protection of GRP pipework and tanks from freezing

Temperature controlled vessels for chemical and industrial processing

GRP immersion heater elements

CHEMICAL RESISTANCE

GRP is widely used in chemical engineering because of its superior chemical resistance in comparison with metals. However, in certain environments, particularly strong concentrated acids and bases, GRP suffers environmental stress corrosion. Several studies have shown that environmental stress failure occurs when micro cracking of the resin allows the corrosive chemical to attack the glass fibres [Hogg and Hull (6), Jones, Rock and Wheatley (7)]. Carbon fibre has excellent resistance to many of the chemicals which attack glass, including hydroflouric acid [Klinger and Barker (8)].

A thin layer of carbon fibre on the surface of GRP vessels exposed to severe environmental conditions provides an extra barrier against stress corrosion failure.

STRUCTURAL APPLICATIONS

The structural properties of composites from carbon fibre papers have not yet been investigated. However, the nature of the papers suggests some interesting possibilities.

One existing structural use of carbon fibre papers is to provide increased stiffness to very lightweight foam products where the lightest woven products would be too heavy.

The major problem in the use of papers for structural applications is that the fibre volume fraction which can be achieved is lower than that obtainable using unidirectional or woven fabrics; the maximum is likely to be around 40%.

The main applications for papers will be those in which the truly isotropic nature of the fabric is used. Physical properties of composites produced from paper are equal in all directions. The use of papers to produce isotropic composites offers significant simplification when compared with the production of quasi-isotropic laminates requiring the placement of reinforcement in four or more directions.

Papers may be used in conjunction with woven or unidirectional fabrics as a load spreading layer to provide a certain amount of reinforcement in directions other than in the main load bearing directions.

CONCLUSIONS

Papermaking is an inexpensive and versatile technique of fibre processing which provides a new dimension to the range of carbon fibre materials. A number of novel applications already exist mainly concerned with the modification of surface properties of reinforced plastics. New applications will undoubtedly appear as the material becomes better known and understood. If carbon fibre is to enter the wider market for industrial and commodity products, then not only must the relative cost of the raw fibre decrease but low cost, high volume techniques for the production of usable intermediates must be found - papermaking is one such technique.

ACKNOWLEDGEMENTS

The author would like to thank the following for their help and advice in the preparation of this paper:

Mr N C Johnson, Fibertec Division, James Cropper PLC

Mr B D Baines, Matthey Rustenberg Refiners

Mr R Stevens, Cambrook Woods

Mr D Hurst, TBA Electroconductive Products

REFERENCES

1. Ministry of Defence (UK) - Def Stan 59-98 "Handling Procedures for Static Sensitive Devices"

2. US Department of Defense - DOD - HDBK-263 "Electrostatic Discharge Control Handbook for Protection of Electrical and Electronic Parts Assemblies and Equipment"

3. Bigg,D.M., and Stutz,D.E., Polymer Composites 4 (1)40 (1983)

4. F.C.C. Rules and Regulations, Vol 11, Part 15 (July 1981)

5. Reilly,J.J., "Solving the Plastics RF Shielding Problem; Practical Approaches", EMI/RFI Shielding Plastics Conference, SPA Chicago section Electrical and Electronic division (June 1982)

6. Hogg,P.J. and Hull,D., "Role of Matrix Properties on the Stress Corrosion of GRP", Proc of Reinforced Plastics Congress (Brighton 1982), (British Plastics Federation) Paper 29

7. Jones,F.R., Rock,J.W. and Wheatley,A.R., Composites 14 (1983) 262

8. Kliger,H.S. and Barker,E.R., "A Comparative Study of the Corrosion Resistance of Carbon and Glass Fibres", 39th Annual Conference, Reinforced Plastics/Composites Institute SPI (Jan 1984)

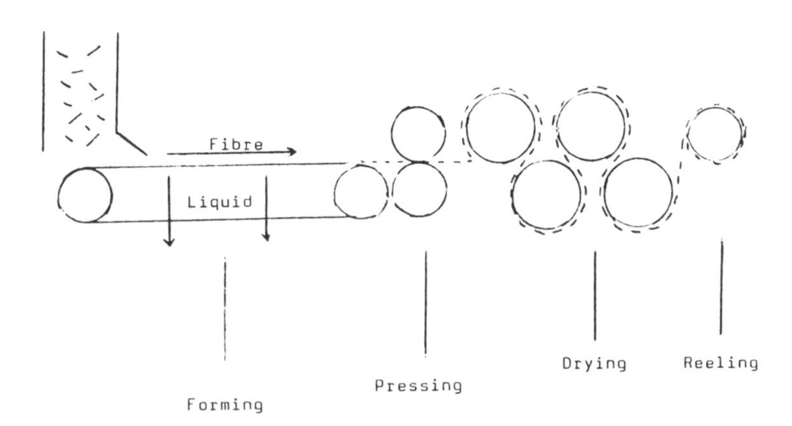

Figure 1 Papermaking System

Figure 2 SEM micrograph of carbon fibre paper x 320

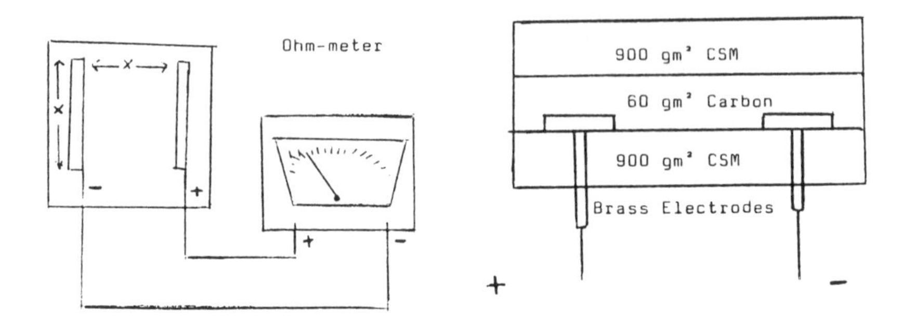

Figure 3 Surface resistivity test equipment

Figure 5 Structure of FRP heating panel

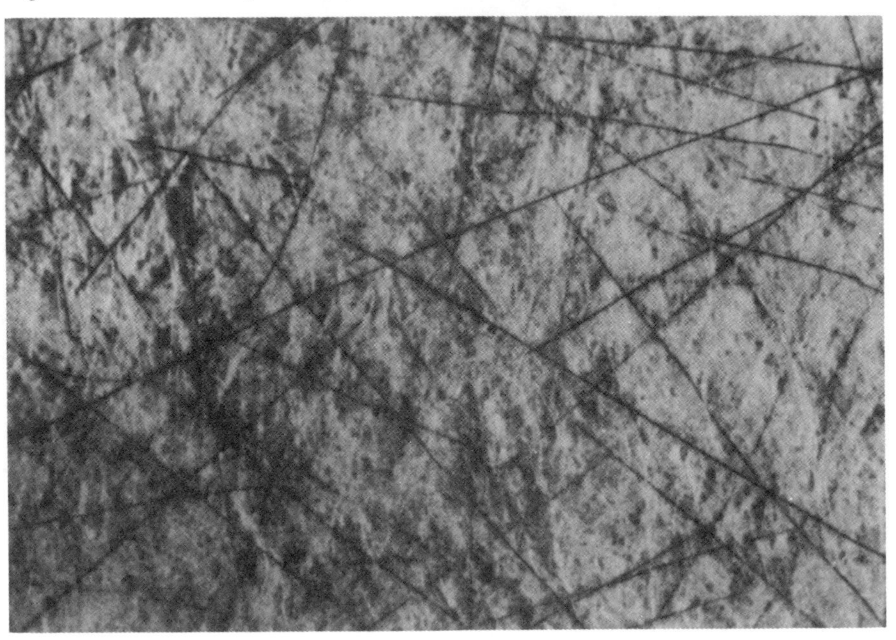

Figure 4 Micrograph of carbon fibre cellulose hybrid x 38

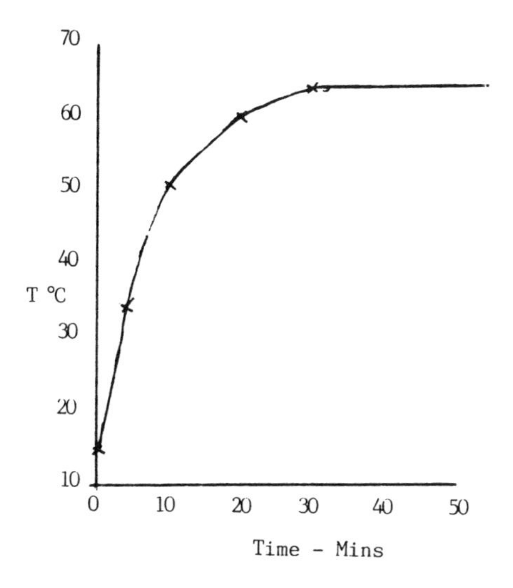

Figure 6 Graph of temperature vs time for heating panel at 12 volts, 1A

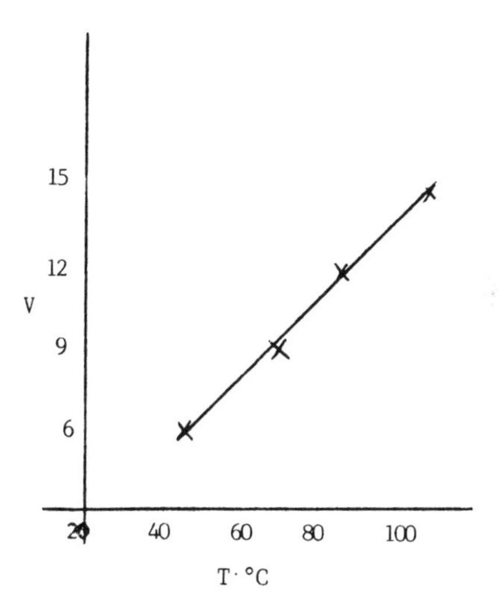

Figure 7 Graph of temperature vs applied voltage for heating panel at constant 1A current

METAL COATED GRAPHITE FIBERS FOR CONDUCTIVE COMPOSITES

Bruce A. Luxon, Ph.D.

American Cyanamid Co.
Metal Coated Fibers Department
Wallingford, CT

Metal coated graphite fibers are described along with some of their applications. Each individual filament consists of an aerospace-grade carbon fiber core surrounded by a plated sheath of metallurgically pure nickel. Composites are described in both thermoplastic and thermoset resin matrices. Specific applications discussed include EMI shielding and electrostatic dissipation for the protection of computers and lightning strike protection of composite aerospace components.

INTRODUCTION

Metal Coated Graphite fibers (MCG) are a new and unique composite material. They combine the high electrical and thermal conductivity of metals with the exceptionally high strength to weight ratio of an aerospace-grade carbon fiber reinforcement. Characterized by a thin, continuous sheath of electroplated metal only 0.4 microns thick uniformly surrounding a 7.0 micron diameter carbon fiber core, MCG fibers retain the remarkable reinforcing ability of the carbon fiber while adding the superior thermal and electrical conductivity of the metal skin. This provides designers and engineers with combinations of properties heretofore unavailable in a single reinforcement

While virtually any electroplatable metal can be applied, the metal of choice is usually nickel for a variety of practical reasons. Nickel is relatively inexpensive and corrosion resistant, while its thermal and electrical conductivity is excellent.

Nickel Coated Graphite (NCG) fibers are produced continuously by a proprietary electroplating process. Similar to its carbon fiber core material, NCG fiber in its most basic form is produced as a continuous bundle (tow) of 3,000 to 12,000 filaments depending on the carbon fiber starting material. Continuous NCG fiber tow is most suitable for weaving into fabric (e.g. 3K plain weave), making impregnated tapes, or filament winding. Also like its carbon fiber parent, NCG fiber is available in a wide variety of secondary forms such as chopped fibers (3 - 25 millimeters long) and non-woven mats. Because NCG fiber is handled just like carbon fiber and is available in the same product forms, it is readily incorporated into existing composite manufacturing schemes with no modifications necessary. This can have significant favorable impact on the produceability of a potential or current application.

Comprised of about 24% by volume nickel, NCG fibers are about 38% less dense than steel with a specific gravity of 3.0 which is about 10% greater than aluminum. Typical fiber properties are listed in Table I with a tensile strength of around 450 ksi and tensile modulus of about 34 msi.

TABLE ONE

FIBER PROPERTIES

PROPERTY	VALUE[1]
Diameter	7.8 microns
Metal Coating Thickness	.4 microns
Specific Gravity	3.0
Tensile Strength	450 ksi
Tensile Modulus	32 msi
Elongation	1.4%
Electrical Conductivity	7 micro OHM-cm(2)
Thermal Conductivity	640 watts/m/^{o}C (3)
Coef. Thermal Expansion	-0.8 10^{-6}/ini/in/^{o}C

1. Nominal
2. as nickel
3. 145% of carbon fiber

These properties are of course largely determined by the carbon fiber core material and are at least as good as those predicted by the rule of mixtures. Material yield for the approximately 7.8 micron diameter NCG fiber with 0.4 micron nickel plating is about 6240 kilometers/kilogram.

NCG FIBER APPLICATIONS

Nickel Coated Graphite fibers are successfully being introduced into a variety of applications where its unique combination of electrical conductivity, high aspect ratio and structural reinforcement provides innovative solutions to demanding problems. In fields as diverse as computers and aerospace composites, NCG fibers are providing physical strength and protection from electromagnetic environmental effects.

Electromagnetic interference shielding

The tremendous growth in the application of engineering thermoplastics as housings for the computer and business equipment industry is a testament to their utility and broad appeal. Over a period of about 15 years they have captured virtually the entire commercial market. Engineering thermoplastics provide design engineers with combinations of aesthetics, physical strength, lightweight, and design flexibility which allows for the integration of form and function impossible to achieve in any other medium.

Recent FCC (U.S.A.) and VDE (F.D.R.) regulations placing limits on electromagnetic interference (EMI) have highlighted perhaps the one thing plastic computer housings cannot do for in their native state plastics are completely transparent to radio frequency signals. To remedy this deficiency, there has been a tremendous surge lately in efforts to produce injection moldable (IM) composites of thermoplastic resins and conductive fillers (e.g. carbon fibers, aluminum flakes, stainless steel fibers, etc.). At best these materials have been only marginally successful. Typically this has been because filler load levels sufficient to provide adequate EMI shielding effectiveness (SE) have not met commercial requirements for reproducible and cost-effective injection molding.

Just introduced, DRY-BLEND* NCG chopped fibers have significantly advanced the state of the art in both EMI shielding effectiveness and produceability.

Designed to be simply tumbled with resin pellets immediately prior to the molding operations, DRY-BLEND* NCG fibers provide upwards of 50 dB of EMI shielding (i.e. 99.999% attenuation of the incident RF signal) with molding behavior that is often indistinguishable from the unfilled resin.

NCG Fiber Loading	Shielding Effectiveness[2]
10% (wt)	25 - 45 dB
15%	35 - 55 dB
20%	45 - 65 dB

1. DRY-BLEND* dB-100 Nickel Coated Graphite fiber in polycarbonate resin.

2. ASTM ES7-83 dual chamber method at frequencies from 30 MHz to 1000
MHz (typical values.

*Trademark American Cyanamid Company

Whereas DRY-BLEND* NCG fiber stresses optimum moldability and cost-effectiveness, CYCOM* NCG fiber is a somewhat different chopped-fiber product designed for the most demanding applications. Here the full structural reinforcement potential of NCG is exploited in combination with other additives and fillers by melt blending them together in a compounding extruder to produce composite resin pellets suitable for injection molding. These materials have demonstrated tensile strengths over 20,000 psi with EMI shielding effectiveness of 95 dB.

PROPERTY	NCG FIBER LOADING (WT. %)			
Tensile Strength (psi)	0	10	15	20
Elongation (%)	9,000	10,000	12,000	14,000
Flexural Strength (psi)	13,200	15,000	16,000	18,000
Flexural Modulus (psi)	325,000	640,000	970,000	1,200,000
HDT @ 264 psi (°F)	270	280	290	300

Common to either of the chopped NCG fiber products are their abilities to import dimensional stability, decrease molding cycle times (hence productivity) due to enhanced thermal efficiency, and they are regrindable with no decrease in properties when the resin manufacturer guidelines are followed.

Electrostatic dissipation

Another fast emerging NCG fiber application includes electro-static dissipation (ESD) for the protection of electronic microcircuitry from catastrophic static discharge. Particularly useful in electronics material handling situation where ICs are used, NCG fiber loadings as low as 7% (weight) will safely dissipate 5KV static charges in .01 seconds or less (regardless of the ambient relative humidity) or more than adequately protecting the assembly station.

Advanced composites

Aerospace Carbon fibers are revolutionizing the aerospace industry with high performance advanced composites being integrated into virtually every component area. Carbon fiber composites provide the equivalent strength of metals at substantially less weight. This weight savings translates into increased range, increased payload, or reduced deployment costs.

As carbon fiber composites become more extensively used in primary structures and leading and trailing edges of aerospace vehicles, reliability in withstanding direct lightning strikes becomes a serious issue. Metal skins and airframes provide efficient electrical continuity participating in a lightning strike as a reliable ground path. Carbon fiber composites however, lacking the inherent electrical conductivity of metal cannot dissipate the high (200,000 Amp) currents imposed by a zone 1 lightning strike whithin the duration of the strike. This results in arcing and even catastrophic failure of the structure. This is especially dangerous when it occurs near fuel containing sections of the craft (e.g. wings, etc.). Arc-over in these areas can ignite the fuel-air mixture in the partially depleted fuel tank.

Nickel coated graphite fiber containing laminates can solve this problem. Lab tests have shown that composite panels containing NCG fiber fabric as the top ply successfully withstand simulated zone 1 lightning strikes. Because each individual NCG filament is completely plated with nickel metal, a 3K plain weave fabric provides about 74 square meters of conductive surface area per square meter of fabric. Thus, NCG fabrics can replace the existing top-most ply of carbon fiber in a laminate thereby importing lightning strike protection and conserving physical properties at a minimum increase in weight.

	ALL CF	1 ply NCG	2 ply NCG	
Tensile Strength (ksi)	109.5	103.5	104.4	rt/dry
Tensile Modulus (msi)	9.1	9.1	9.4	rt/dry
Compressive Strength (ksi)	115.9	103.8	94.0	rt dry
Compressive Strength (ksi)	96.2	94.6	94.2	160°F/dry
Compressive Strength (ksi)	112.3	101.9	81.7	rt/wet
Compressive Strength (ksi)	83.8	84.3	76.3	160°F/wet
Short Beam Shear Strength (ksi)	12.3	11.0	11.7	rt/dry
Short Beam Shear Strength (ksi)	9.7	9.9	9.6	160°F/dry
Short Beam Shear Strength (ksi)	12.2	10.4	9.3	rt/14 day wet
Short Beam Shear Strength (ksi)	8.6	8.4	6.9	160°F/14 day wet

In some cases both the top and bottom plies of carbon fiber will be replaced to preserve the symmetry of the composite. In practice, this is readily accomplished because NCG fabrics and tapes are compatible with all current and emerging composite manufacturing practices. It is handled identically to carbon fiber fabrics and tapes.

Composite hot tools Composite manufacturing is relying more and more on composite tools to fabricate carbon fiber laminate parts. Aside from the obvious cost benefits over expensive matched metal dies (especially for short runs) composite tools provide closer tolerances by matching the thermal expansion characteristics of the tool to the component itself.

NCG fibers offer a number of significant innovations to the composite tooling concept. NCG fibers allow a tool to be integrally heated using the nickel coated graphite fabric itself as the heating element. Resistive heating allows fast heat-up rates of up to 60°F per minute while the 45% increase in thermal conductivity of NCG fiber over bare carbon fiber allows shorter cooling down times for optimum cycle times. Because the NCG fabric itself is heating up, thermal efficiency is very high concentrating the heating at the mold face into the part to be cured. Energy and time are not wasted heating up the surrounding environment needlessly. Being oxidatively more stable than carbon fiber alone, NCG tools should have excellent life expectancies. Microprocessor controls allow for very precise temperature distributions and thermal ramps.

CONCLUSION

Nickel coated graphite fiber have been presented as an innovative approach to applications requiring conductive reinforcement. Obviously, the applications cited are not an exhaustive list. The clever reader will think of many others where the high strength to weight of carbon fiber combined with the electrical conductivity of metals can be used to advance his technology.

Bruce A Luxon is currently the Technical Services Manager for the Metal Coated Fibers Department of the American Cyanamid Company located in Stamford, Connecticut U.S.A. Prior to joining American Cyanamid in 1983, Dr Luxon worked as an Advanced Development Chemist for the General Electric Co. and as a consultant to the Illinois Tool Works Co. He received his PhD. in Physical Chemistry from the University of Illinois in 1977. Dr Luxon has publications in the areas of quantum chemistry and high-field nuclear magnetic resonance. His patents range from miniature electrical devices to conductive resins and composites.

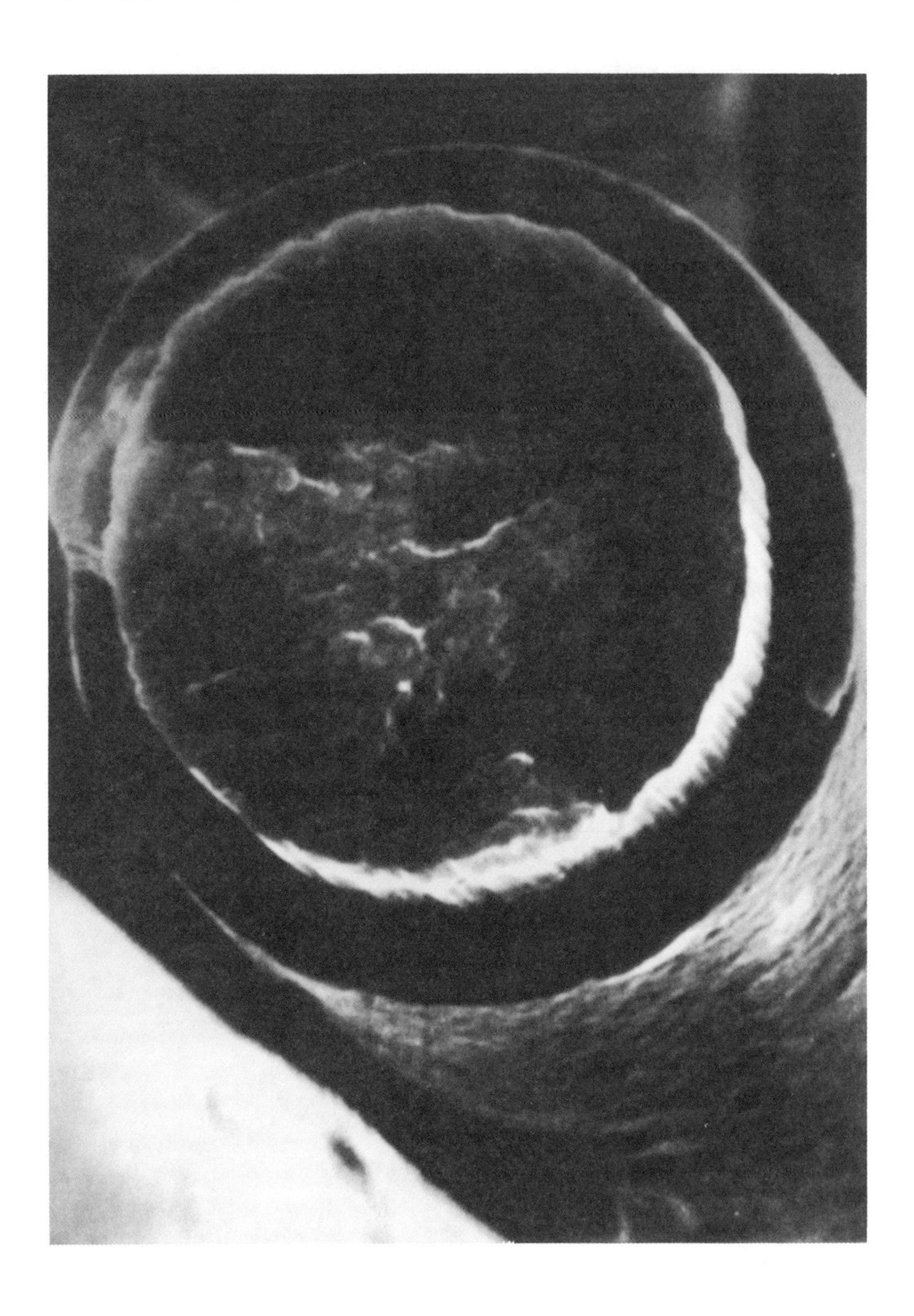

CFRP MOULDING COMPOUNDS

(Speaker from LNP Corporation)

Paper not available at press time.

PROGRESS AND PROSPECTS IN CFRTP'S

B. Ballance

Wilson-Fiberfil International

CFRTP's have become a significant factor in the engineering plastics market. As alternatives to metals for precision parts, better design use and more varied options on materials have stimulated growth. 3 Main criteria point to the selection of CFRTP's but performance is profoundly affected by the actual fibre used. Cost/performance may be improved by customising materials to design requirements and by carefully controlling moulding aspects. Correct usage of existing options can be expected to meet more demand than radically new developments.

INTRODUCTION

Carbon Fibre reinforced thermoplastics (CFRTP's) are a speciality sector of the much larger thermoplastics industry involved with injection moulding. Being part of the industry, it shares similarities to it, but differs from it in many ways. A simple example, thermoplastics are normally moulded economically at annual requirement levels of 20-40 thousand parts and upwards. CFRTP's can be economical for run sizes of only a couple of hundred, quite a difference.

CFRTP's joined this industry as an expensive reinforcement option - 10 times or more the equivalent glass reinforced type - and offering distinctive properties (1,2,3). Five years ago, one analysis (4) predicted an annual market of around 500 tons of compound by the mid-80's ; from a starting point of only a few tens of tons. In the manner of most forecasts, it has turned out wrong, an underestimate by a factor of at least 50 %. The growth, too, has come from unexpected areas and this paper will review some of the factors which have brought that about, and examine the present driving forces that may direct its future.

KEY POINTS OF RECENT TRENDS

Over the past decade, two strong influences have been developing and have contributed to the present state of the market.

A better understanding of the composites by materials engineers and designers is the first of these. Extensive promotion, the growth in use, the approach of the specialist supplier of "designing" materials, have all increased awareness of CFRTP's possibilities through all industries. Consequently usage has not simply increased in a core number of materials, but a tremendous variety of grades has been spawned (Table 1), each new grade being a solution to a past problem.

Different loading levels - steps of 5 or 10% - can be added to by the use of further functional additives ; ptfe, silicones, molybdenum disulphide, graphite, glass fibre ; each contributing something new - the result is a bewildering mass of grades. Often these additives are included in parts requiring built-in or auto-lubrication properties. Carbon fibre itself demonstrates improvements in this direction. Its lower surface frictional characteristics, lack of abrasiveness and increased heat dissipation (Table 2) lead to composites with improved tribological properties - lower wear on part and the mating surface, lower friction (Table 3) and higher PV limits.

Improved cost/performance is the second influence. This has been achieved by better performance through i) improvements in the fibres (better fibre/polymer bonding) - resulting in about 15 % better composite quality
ii) better use of those fibres - compounding techniques have largely come to terms with carbon fibres and now yield more consistent and better quality ; moulding techniques have been addressed and cumulative production experience has brought us some way down the learning curve. As a caution, it should be noted that CFRTP's represent a minute fraction of the thermoplastic market and this good experience is limited to a relatively small number of technical moulders.

Together with improved performance has come a reduction in cost, both real and relative. In actual money terms, there has been a decline in composite prices of about 30 % and this at a time when competitive forms have been increasing. As a result, the early 10:1 plus ratio is now 6 or 8:1 and on the newer higher performance materials eg PES, PEI it may only be 3 or 4:1.
This comparison masks the fact that CFRTP's are not in competition with natural or glass reinforced plastics but metals. With strength to weight ratios comparable to metals (Table 4), added to the design flexibility of injection moulding - increasing stiffness by design complexities - the inherent low shrinkage and stability of CFRTP's ; these all add up to a low cost route to precision intricate technical parts. And in the metals industry, cost inflation in labour and machining time have now given CFRTP's a real edge.

Pitch

Pitch fibres are often encountered in the literature but seldom these days in CFRTP's. The type in use was a shorter and cheaper form which for a period of 5-8 years provided a cost effective alternative to PAN fibres in certain instances. Those particular advantages have been lost and on both technical and cost grounds there is little justification for its use.
In the cyclical pattern of the plastic industry, pitch fibres will no doubt return - the many announced projects in Japan, USA and Europe bear witness to that - with the fibres mutated perhaps to be equivalent or superior to PAN. This would be something new in CFRTP's not comparable with the earlier types and would doubtless further the evolution of the CFRTP species.

THE PRESENT STATE OF THE ART

The applicational science of CFRTP's is largely an empirical one, not a theoretical science and sometimes borders on being an art.

From observed practice, materials are constructed to solve perceived problems - the starting point is what is known about carbon fibre in thermoplastics and its ability to solve real problems in industry. Carbon fibre may be used for any of 3 main reasons.

i) the superior reinforcing additive : this manifests itself in different forms depending on need but can refer to modulus, strength to weight, creep and dimensional stability, fatigue or chemical resistance.

ii) Conductivity - in comparison with glass and most other fibres it is electrically conductive.

iii) Lubricating additive - extends part life by lowering wear and friction and raising PV values.

Comparison of different fibres

In each of the above modes, composite performance depends on fibre length, affected by both compounding and moulding ; the fibre polymer interface, namely sizing technology, and intrinsic fibre properties (high strain or stress or modulus etc..).

Differences in the last of these are small (Table 5) and overshadowed by the invisible importance of the first two (Table 6). These short term laboratory tests show wide differences in a controlled compounding and moulding study (5). Impact strengths, the property most directly dependent on the fibre - polymer mechanical interaction, are significantly different.
In comparing data from different sources, a correlation of better than 10-15% on many properties is all that can be hoped for, on impact it is much worse, assuming the same method is even used. Consequently, published data alone is insufficient to judge a material. To illustrate further, the measured changes between a long and short fibre are not significant except on impact with its attendant reservations. Yet in another context it has been found to be quite superior in respect to dynamic response to load (6) (Figure 1) and unpublished studies show considerable advantages at higher temperatures.

The conclusion is that as a reinforcer, some fibres perform better. However faced with this situation of material differences obscured even to the initiated, it is easy to perceive how materials, to their detriment, are viewed as generic types differentiated only on price. In the arena of premium performance and price materials this is a trap for the unwary.

Conductivity in CFRTP's

The demand for environmentally safe materials from such dissimilar worlds as the microchip, mining or handling of explosive media, has made conductivity a key property in carbon fibre's portfolio. From loadings of as little as 10 % by weight a material is made static dissipating, protecting electronic circuits or avoiding spark generation. Conductivity then increases with higher loadings with a more linear relationship than many other additives (7). This allows for a broader spread of composite conductivities, controlled to an extent to meet end-use requirements for example

for static dissipating	$10^4 - 10^6\ \Omega.cm$
functional elements in high impedence circuits	$10^2 - 10^3\ \Omega.cm$
shielding from radio frequency interference	$10^0 - 10^2\ \Omega.cm$

This last is a complex subject in itself but it suffices to mention that r.f. screening requirements are specific to individual applications and may in some case need to be only 10 dB (90% screen) or as much as 80 dB (99,999999 % screen). At lower screening levels CFRTP's can provide a cost effective solution where a particular balance of properties is called for.

Matching materials to requirements

Not surprisingly, a multifunctional additive such as carbon fibre combined with the diverse possibilities of thermoplastics can find itself satisfying different needs with each new application.

The "customising" role played by the specialised supplier of CFRTP has allowed these nuances to be met more effectively, through correct selection of materials and fibres, by special additives or mixtures, by optimising the materials moulding characteristics and by participating in design and tooling. This diversity of supply has been the big spur to recent growth and highlights the way forward - building materials to perform : a matching of materials to requirements using in-depth, informed and objective knowledge of the industry.

The subtlety with which materials can be differentiated is illustrated in the next sequence of tables ; each grade reflecting small changes to suit a particular need (Tables 7-10) in the area of conductivity.

Table 7 shows a progression of conductivities from a familiar engineering material as starting point ; a 20 % glass reinforced polycarbonate. Mechanical properties can be preserved within a limit of conductivity by hybrid formulation - at higher levels property trade-offs appear.

Table 8 repeats part of the sequence on a modified PPO.

A third engineering "workhorse" is illustrated in Table 9. 30% glass reinforced nylons account for over 50% of all fibre reinforced thermoplastics and the optional trade-offs to ESD in this deliberately limited choice are clear, impact or tensile.

Finally in Table 10, the carbon fibre modification of ABS is illustrated in two modes ESD or r.f. screening.

The benefits of "customising" can be shown just as strongly in both the areas of pure reinforcement and self lubrication.

FUTURE TRENDS

Progress and growth will continue with CFRTP's, continuing past trends and perhaps adding some new elements. More grades will be tailored to satisfy special demands and this process will yield more than any particular new fibre development. That said, it could be expected that binder systems will improve, making more out of the fibres we have.

A comparison of glass and carbon fibre (Table 11) at the same weight percent addition seems to suggest that glass fibres are proportionally a more efficient fibre. This may be partly due to carbons' lower elongation or partly to inferior load transfer mechanisms. However improvements will aid the future prospects for the industry but not be a precondition to it.

More substantive developments are likely to occur from lengthened fibres in the end product. Firstly from the typical 300-400μ to a few millimetres ; still discreet chopped fibres and pellets of moulding compounds.
This established product option is being re-evaluated in the light of some technical improvements.

Further lengthening to prepregs is an important development for thermoplastics but is outside the scope of this paper.

The moulding industry also has a big role to play.
Improved tooling and moulding practices will enable more projects to be realised by delivering better parts. Published work on moulding of CFRTP's is limited (2, 8) and not much of eg Cadcam data bases refers to these materials. Consequently experience will be important, coming both from the technical moulder and his specialist supplier.

In summary, the CFRTP sector is in a robust and dynamic state.
Properly nourished with the right degree of interest from the user, its current trends can confidently be expected to continue.

References

1. Carbon Fibre Reinforced Thermoplastics, A.P. Cluley, Sampe London 1983.
2. Carbon-reinforced Thermoplastics, S.R. Gerteisen & S.D. Gerbig, Plastics Engineering January 1983.
3. Wear Characteristics of Carbon Fibre Reinforced Thermoplastics, J. Theberge, B. Arkles, Asle, 29th annual meeting Cleveland, O. April 28-May 2, 1974.
4. Carbon Fibre Reinforced Thermoplastics, Trewin, Turner & Cluley, ICCM-3 Paris 1980.
5. The Evaluation of Carbon Fibres for Optimum Reinforced Thermoplastic Properties, S.R. Gerteisen, SPE 43rd Annual Tech. Conf.
6. Composite Materials in Handprostheses, U. Persson, ISPO, IV World Congress Sept. 83.
7. Thermoplastic Composites for Electrostatic Dissipation and EMI Attenuation, J.E. Travis, 39th Annual Conference, Reinforced Plastics/Composites Institute. SPE January 1984.
8. Carbon Fibres add Muscle to Plastics, J. Theberge, B. Arkles, R. Robinson, Machine Design Feb. 1974.

Bruce Ballance is the European Marketing Specialist at Wilson-Fiberfil International, having joined the company in 1982. Mr Ballance is a graduate of Leeds University in Textile Physics and worked for some years in the synthetic fibre industry before moving into the plastics industry where he has some 11 years experience in marketing and technical roles with several firms. He is a member of the Society of Plastics Engineers and the British Institute of Management and lives in Namur, Belgium.

TABLE I

AVAILABILITY OF PAN CARBON FIBRE REINFORCED THERMOPLASTICS

Polymers	Max. loading % wt	Features
General Purpose		
PP - ABS - mod. PPO/PPE - TPU	30-40	
Engineering		
Polyamides	50	best property balance for general use
Polyesters	50	moisture resistant
Polycarbonate	50	stability, flame resist.
Acetal	40	
High temperature and flame retardant		
PPS	50	solvent resistance
PES - PSu - PEI	40	stability, very low shrinkage
PEEK	40	up to 260°C temp. resist.
Fluoropolymers		
Pvdf - Etfe - FEP	30	chemical resistance

TABLE II

TEMPERATURE AT BEARING SURFACE[1] ON REINFORCED PPS

Composite	Measured Temperature
PPS + Glass + lubricants	150 - 180°C
PPS + Carbon Fibre + lubricants	50 - 70°C

1 load of 5-7 MPa at up to 100 m/min. velocity.

TABLE III

EFFECT OF CARBON FIBRE ON TRIBOLOGICAL PROPERTIES OF POLYAMIDE 6.6

	Limiting PV at		Wear factor	Coeff. of friction	
	30 m/min	300 m/min	10^{-8} mm^2/kg	static	dyn.
Polyamide 6.6	5.26	4.2	80	0.20	0.28
6.6 + Ptfe	35.8	16.8	4.8	0.10	0.18
6.6 + CF + Ptfe	90.5	42	4	0.11	0.15

TABLE IV

PROPERTIES OF METALS AND REINFORCED PLASTICS

Property	Units	METALS				Polyamide 6.6		
		Cast Steel	Die cast alloys					
			Al	Mg	Zn	Unfilled	30% glass	30%CF
Specific gravity		8	2,8	1,8	6,6	1,16	1,39	1,28
Tensile Strength	MPa	500	200	227	283	80	180	250
Specific Strength	$MPa/g.cm^{-3}$	62	70	126	43	69	129	195
Flexural Modulus	GPa	210	69	47	41	2,8	9	18,6
Specific Modulus	$GPa/g.cm^{-3}$	26	25	17-26	7-15	2,4	6,5	14,5
Tensile Elongation		25	8			10	3-4	3-4
Coefficient of thermal expansion	$10^{-6}/K$	14	22	25	27	80	24	10

TABLE V

PROPERTIES OF 30% CARBON FIBRE REINFORCED POLYAMIDE 6.6

Fibre		standard chopped	long	high strain	5 mu Ø
Tensile Strength	MPa	250	280	275 (+11%)	275 (+11%)
Flex. Strength	MPa	360	372	385 (+7%)	390 (+8%)
Flex. Modulus	GPa	18.6	21.4	20.7 (+11%)	20.7 (+11%)
Notched Izod	J/m	85	140	85	97 (+14%)
Surface Resistivity		500	30	250	250

TABLE VI

PROPERTIES OF 30% WEIGHT PAN CARBON FIBRES IN POLYAMIDE 6.6

Properties		Fibre Suppliers I	II	III	IV
Tensile Strength	MPa	245	240	210	224
Flexural Modulus	GPa	18,2	18,0	17,2	15.8
Flexural Strength	MPa	330	350	288	322
H.D.T.	°C	254	254	255	254
Surface Resist.	Ω	75	100	30	100
Notched Izod Impact	J/m	76	81	54	65

TABLE VI (a)

PROPERTIES OF 30 % WEIGHT PAN CARBON FIBRES IN PPS

Properties		Fibre Suppliers I	II	III
Tensile Strength	MPa	186	186	150
Flexural Strength	MPa	245	250	224
Flexural Modulus	GPa	23,5	23,4	21,4
Izod Impact	J/m	54	80	60
HDT	°C	260	262	260
Surface Resist.	Ω	200	20000	500

TABLE VII

PROPERTIES OF REINFORCED STATIC DISSIPATING/CONDUCTIVE POLYCARBONATES

			ESD			Conductive
		20% GF	#1	#2	#3	#4
Tensile Strength	MPa	107	120	110	115	120
Elongation	%	3.0	4.0	3.0	2.4	3.5
Flexural Strength	MPa	151	175	165	172	172
Flexural Modulus	GPa	5.5	7.0	7.6	7.6	8.8
Notched Izod	J/m	134	113	90	92	75
Specific Gravity		1.34	1.31	1.24	1.27	1.26
HDT	°C	143	142	142	143	142
Vol. Resistivity	Ω.cm	10^{15}	10^{7}	10^{5}	10^{4}	200

TABLE VIII

PROPERTIES OF ESD/CONDUCTIVE MODIFIED PPO

		30 % GF	GF/CF
Tensile Strength	MPa	96	124
Notched Izod	J/m	95	90
Specific Gravity		1.31	1.20
HDT	°C	137	137
Vol. Resistivity	Ω.cm	10^{15}	10^{4}
Flexural Modulus	GPa	7.6	8.9
p/cm^3		0.30	0.71

TABLE IX

PROPERTIES OF REINFORCED STATIC DISSIPATING POLYAMIDE 6

PROPERTY		J-3/33	+ CF	Impact modified + CF
Tensile Strength	MPa	186	190	140
Elongation	%	4.0	2.5	4.0
Flexural Strength	MPa	276	280	240
Flexural Modulus	GPa	9.0	12.0	10.0
Notched Izod	J/m	140	115	215
Specific Gravity		1.39	1.35	1.31
HDT	°C	257	255	246
Vol. Resist.	Ω.cm	10^{15}	10^{5}	10^{5}
Mould shrinkage	%	0.2	0.2	0.2

TABLE X

PROPERTIES OF ESD/CONDUCTIVE ABS

		Nat.	ESD	Highly Conductive
Tensile Strength	MPa	45	80	110
Notched Izod	J/m	220	60	64
Specific Gravity		1.06	1.09	1.14
HDT	°C	99	102	102
Vol. Resistivity	Ω.cm	10^{15}	10^{4}	10^{0}
Flexural Modulus	GPa	2.4	6.5	12.4
p/cm^3		0.15	0.50	1.03

TABLE XI

PROPERTY COMPARISON OF REINFORCING FIBRES

Fibre		'E' glass	PAN Carbon
Diam. mu		13	8
Tensile Strength	GPa	3.44	3.24
Youngs Modulus	GPa	72.4	234
Elongation		4.8	1.5
Thermal Conductivity (BTU in/hr. ft^2.°F)		7	60
Coeff. of Thermal Expansion 10^{-6}/K		1.6	-0.55
Specific Gravity		2.54	1.77
16% Composite in Polyamide 6.6			
Tensile Strength	MPa	175	200
Flex. modulus	GPa	9.0	13.0
Notched Izod Impact	J/m	108	70
Unnotched Iz. Imp.	J/m	1334	560
HDT	°C	252	254
S.G.		1.39	1.24
Vol. Resistance		10^{15}	10^3

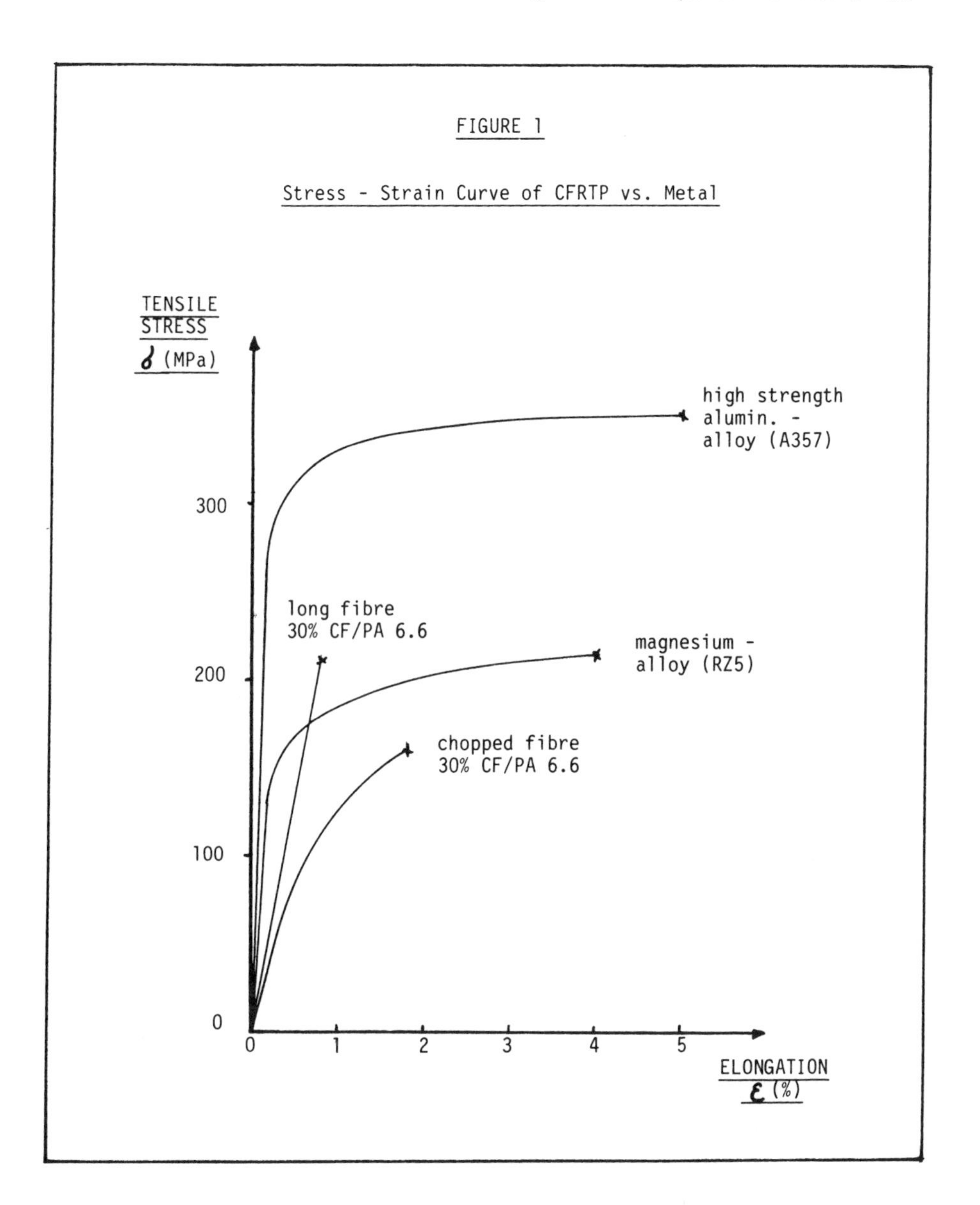
FIGURE 1
Stress - Strain Curve of CFRTP vs. Metal
TENSILE STRESS
σ (MPa)
300
200
100
0
0
1
2
3
4
5
ELONGATION
ε (%)
high strength alumin. - alloy (A357)
long fibre 30% CF/PA 6.6
magnesium - alloy (RZ5)
chopped fibre 30% CF/PA 6.6

CRYSTALLIZATION STUDIES OF THERMOPLASTIC COMPOSITES BY DYNAMIC MECHANICAL ANALYSIS

J.N. Leckenby, D.C. Harget

Du Pont (UK) Ltd.
Wedgwood Way
Stevenage, UK

W.J. Sichina, P.S. Gill

Du Pont Company
Concord Plaza
Wilmington, DE 19898

Characterization of thermoplastic composites, eg PEEK/Carbon fibre composite, is made easier using modern thermal analysis techniques. Comparison of PEEK prepregs is made using Differential Scanning Calorimetry (DSC) and Dynamic Mechanical Analysis (DMA). This paper describes the use of thermal analysis in such characterization techniques as influence of inducing agents on crystallization behaviour, dynamic modulus and loss, transition temperatures and optimization of processing parameters.

INTRODUCTION

Thermoplastic/carbon fibre composite systems have undergone development and evaluation for the past nine years. The aerospace industry has shown interest in themoplastic composites for a number of reasons including major reductions in fabrication costs, improved damage tolerance, improved reprocessibility and repairability, reduced moisture pick-up, and the associated moisture degradation. The processing of thermoplastic composites for aerospace structural components is inherently simpler than its epoxy-based counterpart as the thermoset matrix has three processing variables (chemistry, time, temperature), while the thermoplastic matrix eliminates the chemistry variable.

One of the superior thermoplastics utilized in continous carbon fiber reinforced composites is poly aryl-ether-ketone or PEEK. PEEK offers excellent solvent resistance and good mechanical properties when compared to other thermoplastics. The properties of the final laminate largely depend on the structure of morphology obtained by the semi-crystalline thermoplastic matrix. Factors which can influence the morphology, and hence the mechanical properties, of the PEEK matrix include:

Carbon Fiber Surface Properties
Nucleating Agents
Copolymer(s)
Cool-Down Rate
Time and Temperature in Melt
Molecular Weight

As these factors can strongly influence the structure achieved by the matrix during consolidation, it is essential that the crystallization behavior associated with the PEEK/carbon fiber prepregs be characterized. By understanding the basic morphology of the resin matrix, one can adjust processing parameters during the fabrication process to optimize the mechanical properties of the final product.

Thermal analysis is an ideal technique for the rapid and precise characterization of thermoplastic composite materials. This paper describes the use of thermal analysis in such characterization techniques as influence of inducing agents on crystallization behavior, dynamic modulus and loss, transition temperatures and optimization of processing parameters.

EXPERIMENTAL

The PEEK prepregs analyzed in this study were manufactured by Imperial Chemicals, Inc. (ICI). Test specimens were prepared from single sheets of prepreg using a 5/16 inch paper punch to provide a total sample mass of 6 to 8 mg. The disks were placed in DSC aluminum sample pans, which were then covered with a lid and crimped. A Du Pont 910 Differential Scanning Calorimeter was utilized in conjunction with a 9900 Thermal Analyser. For the Dynamic Mechanical Analysis (DMA) portion of the study, a laminate was prepared by taking 5 strips (30 mm x 12 mm) of prepreg and mounting the stack in the Du Pont 982 DMA equipped with vertical clamps. The plies were clamped to a torque of 110 N-cm. The sample was then heated at 10°C/min to 380°C, held for 5 minutes and cooled at 5°C/min to 100°C. This allowed the PEEK thermoplastic matrix to melt and permitted the formation of a laminate. The clamps were retightened to a torque of 110 N-cm for the heating and cooling experiments. The Du Pont 982 DMA monitors the resonant frequency and damping properties of a sample. The 9900 thermal analyzer with software calculates the tensile storage modulus (E') and loss modulus (E") from the frequency and damping data.

PROPERTIES OF PEEK

PEEK thermoplastic behaves similarly to polyethylene terephthalate (PET) except that the major transitions of PEEK occur 75°C higher than those of PET. Shown in Figure 1 is the thermal curve obtained by heating a sample of amorphous PEEK at 10°C/min in a Du Pont 910 Differential Scanning Calorimeter. The glass transition Tg occurs at 144°C. Upon passing through the glass transition, the molecules acquire sufficient mobility to undergo nucleation and crystallization at 170°C. The melt temperature of PEEK is observed to be 340°C. Shown in Figure 2 is the thermal curve obtained by cooling the sample of PEEK/carbon fiber composite at a rate of 10°C/min. The crystallization behavior of PEEK is analogous to that observed with PET in that it occurs relatively slowly. The slow crystallization means that, depending on the processing conditions, very different properties can be obtained when the material is cooled from the melt.

ISOTHERMAL CRYSTALLIZATION STUDIES

Isothermal crystallization studies are informative providing the temperature at which a given thermoplastic achieves its most rapid rate of crystallization. This is particularly important for PEEK/carbon fiber composites since rapid crystallization of the PEEK matrix is necessary to obtain mechanical properties. (1)

In the isothermal crystallization rate (ICR) studies, a sample of PEEK/carbon fiber composite is heated above its melting point (to 380°C) and is maintained for a period long enough to melt the existing crystalline regions. The molten polymer is then rapidly cooled to the desired temperature below the melting point and the resulting crystallization is monitored. The time at which the crystallization exotherm reaches its maximum is determined. Shown in Figure 3 are the isothermal crystallization curves obtained for APC-2 (Advanced Polymer Composite-2 manufactured by ICI). The samples were maintained at 380°C for 5 minutes prior to cooling down.

As with most thermoplastics, PEEK has a certain temperature regime in which crystallization from the melt proceeds most rapidly. This temperature regime is determined by plotting the maximum crystallization rate time, versus the corresponding isothermal crystallization temperature. This is shown in Figure 4 for the PEEK/carbon fiber material. As may be seen, the PEEK matrix crystallizes most rapidly at a temperature of 230°C.

DYNAMIC MECHANICAL ANALYSIS

The Du Pont Model 982 Dynamic Mechanical Analyzer (DMA) is a powerful technique used in the characterization of composites (2). The DMA detects transitions which cannot be measured by more conventional thermal analytical techniques. One such transition is the glass transition of the semi-crystalline PEEK matrix. Shown in Figure 5 is the DMA thermal curve of a sample of APC-2 which has a crystalline content of 45%. The solid line represents the storage modulus (E'), while the dashed and dash-dot-dashed lines reflect the tensile loss modulus (E") and tan delta (tan o=E"/E') properties of the composite, respectively. The trace reveals that the stiffness or tensile storage modulus of the material remains constant up to 160°C.

At this point, the amorphous regions of the PEEK matrix begin to soften and the resulting glass transition is seen as a sharp decline in E' and a loss modulus peak at 200°C. With further heating, the crystalline regions of the PEEK matrix begin to melt resulting in a further drop in the storage modulus and a peak in the tan o curve at 370°C.

The DMA can be utilized to observe the crystallization of the molten PEEK and thus enchances the results obtained by DSC. Using a new Liquid Nitrogen Cooling Accessory recently developed by Du Pont Instruments, it is possible to achieve uniform cooling rates from 1 to 20°C/min. Shown in Figure 6 are the DMA cooling curves obtained at different cooling rates on a laminate consisting of 5 plies of APC-1. The crystallization of the PEEK matrix is observed as a sudden increase in the stiffness or resonant frequency of the composite material. As the cooling rate is increased, the temperature at which crystallization proceeds is moved toward lower temperatures.

The DMA cooling studies revealed that cooling at a rate of 20°C/min may be too fast in that structural gradients occur across the laminate. The sample that was cooled at 20°C/min had a 20% lower tensile storage modulus at room temperature than the composites cooled at slower rates. It is hypothesized that at 20°C/min, heat was extracted on the surface plies at a far greater rate than in the interior of the laminate. The crystalline contents of the outer layers are significantly lower than the interior portions. This then results in the laminate obtaining a lower overall tensile storage modulus.

APC-2 VERSUS APC-1

The manufacturer of PEEK/carbon fiber prepreg is in the process of improving the properties of the material to ensure the formation of a laminate which does not exhibit structural gradients in the matrix. This can be done by altering the properties of the composite (e.g. lowering the PEEK molecular weight, addition of nucleating agents, changing the characteristics of the carbon fiber surface) to produce a matrix which crystallizes more rapidly. Shown in Figure 7 is a comparison of the stiffness of two PEEK materials when cooled in the DMA at a rate of 5°C/min. The thermogram reveals that the newest generation of PEEK composite (APC-2) crystallizes more readily than its predecessor (APC-1). Thus, it is expected that APC-2 would be less likely to produce structural gradients in a laminate during the consolidation process.

DSC studies also show that APC-2 is much less sensitive towards processing conditions than APC-1. The time and temperature melt conditions strongly affect the crystallization behavior of APC-1 (3). Shown in Figure 8 are the isothermal crystallization curves obtained at 310°C for APC-1 after being maintained in the molten state for various times at 380°C. The crystallization curves show that as the PEEK matrix in APC-1 is held for longer times in the melt, the resulting crystallization exotherms become increasingly more shallow and the times to reach the maximum rate of crystallization become longer.

Shown in Figure 9 are the results of the same experiment conducted on APC-2. These results show that the crystallization of APC-2 is independent of the melt conditions, since increasing the melt time at 380°C does not have an appreciable effect on the shape or position of the crystallization exotherm. These results indicate that there would be much less chance of producing structural gradients in a laminate comprised of APC-2 prepreg.

CONCLUSIONS

The physical properties of PEEK/carbon fiber composites are readily characterized by thermal analytical techniques, particularly Dynamic Mechanical Analysis and Differential Scanning Calorimetry. PEEK behaves similarly to PET in that it crystallizes relatively slowly. The PEEK matrix reaches its maximum crystallization rate at a temperature of 230°C. In order to avoid the formation of structural gradients across a thick laminate, it is essential that the properties of the thermoplastic and carbon fibers can be adjusted to produce a matrix which crystallizes as rapidly as possible. The manufacturer of PEEK/carbon fiber composite materials has achieved success in this endeavor as the latest generation of the composite (APC-2) crystallizes significantly more rapidly than APC-1. In addition, studies reveal that the processing conditions, especially time and temperature in the melt, affect APC-2 much less than APC-1.

REFERENCES

1. D.J. Blundell and B.N. Osborn, "The Morphology of Poly Aryl Ether Ether Ketone", Polymer, 27, 953-958 (1983).

2. John V. Wood, "Thermal Analysis of Electronic Materials", Du Pont Publication E-60874.

3. J.G. Shukla and W.J. Sichina, "Thermal Behavior of Carbon Fiber Reinforced Polyetheretherketone", ANTEC Conference Proceedings, 1984, P. 265-267.

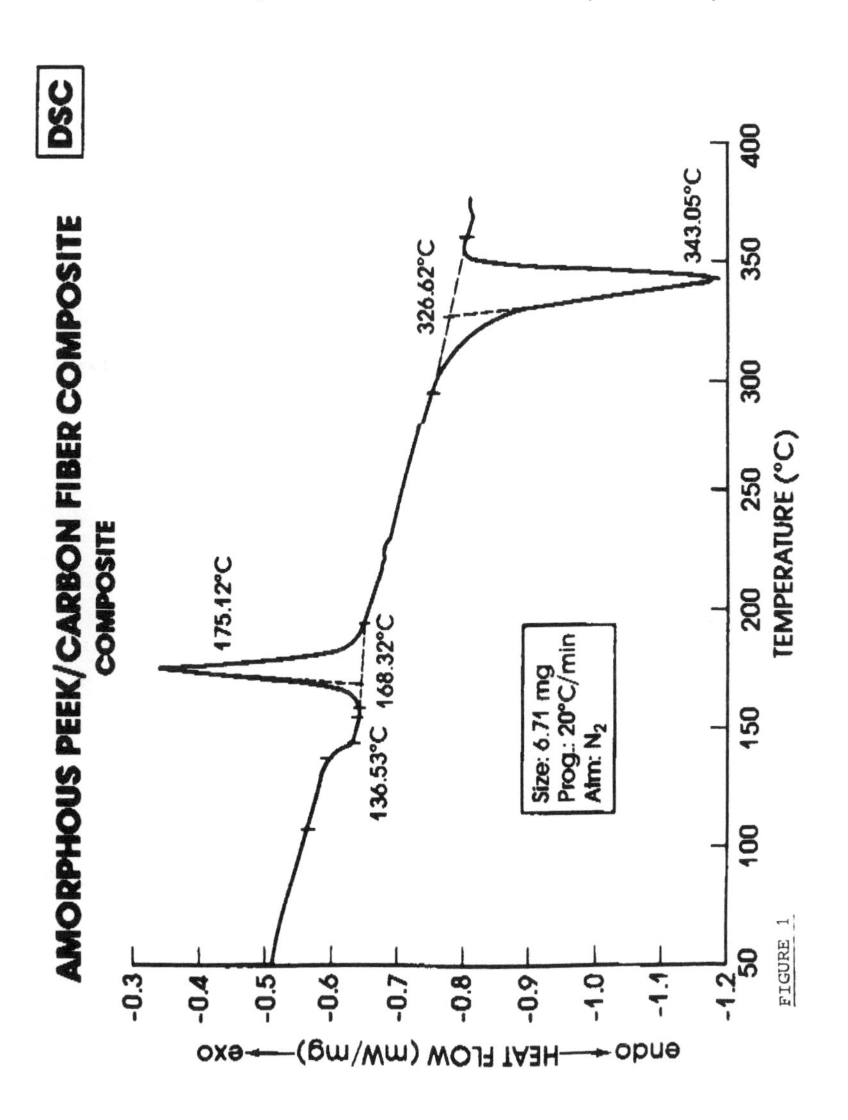
AMORPHOUS PEEK/CARBON FIBER COMPOSITE
COMPOSITE
DSC
175.12°C
136.53°C
168.32°C
326.62°C
343.05°C
Size: 6.71 mg
Prog.: 20°C/min
Atm: N2
endo ← HEAT FLOW (mW/mg) → exo
-0.3
-0.4
-0.5
-0.6
-0.7
-0.8
-0.9
-1.0
-1.1
-1.2
50
100
150
200
250
300
350
400
TEMPERATURE (°C)

FIGURE 1

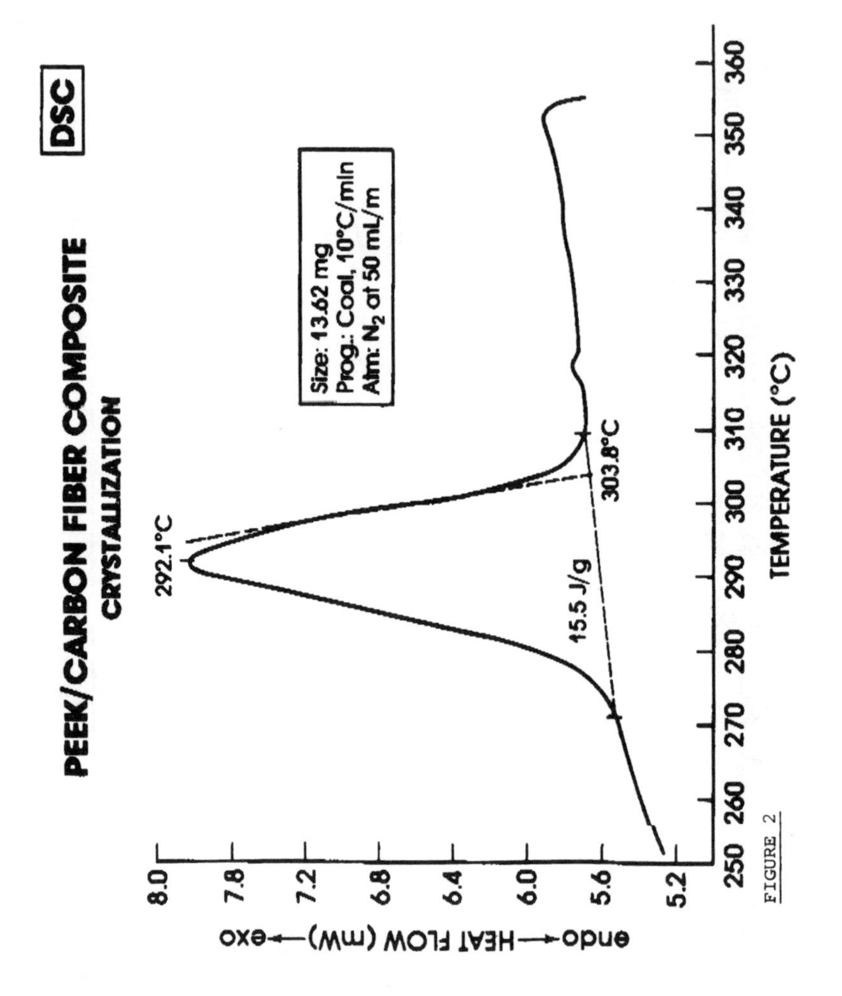
DSC
PEEK/CARBON FIBER COMPOSITE
CRYSTALLIZATION
Size: 13.62 mg
Prog.: Cool, 10°C/min
Atm: N2 at 50 mL/m
292.1°C
303.8°C
15.5 J/g
endo←HEAT FLOW (mW)→exo
8.0
7.8
7.2
6.8
6.4
6.0
5.6
5.2
250
260
270
280
290
300
310
320
330
340
350
360
TEMPERATURE (°C)

FIGURE 2

CRYSTALLIZATION OF PEEK PREPREG
TYPE 2
DSC

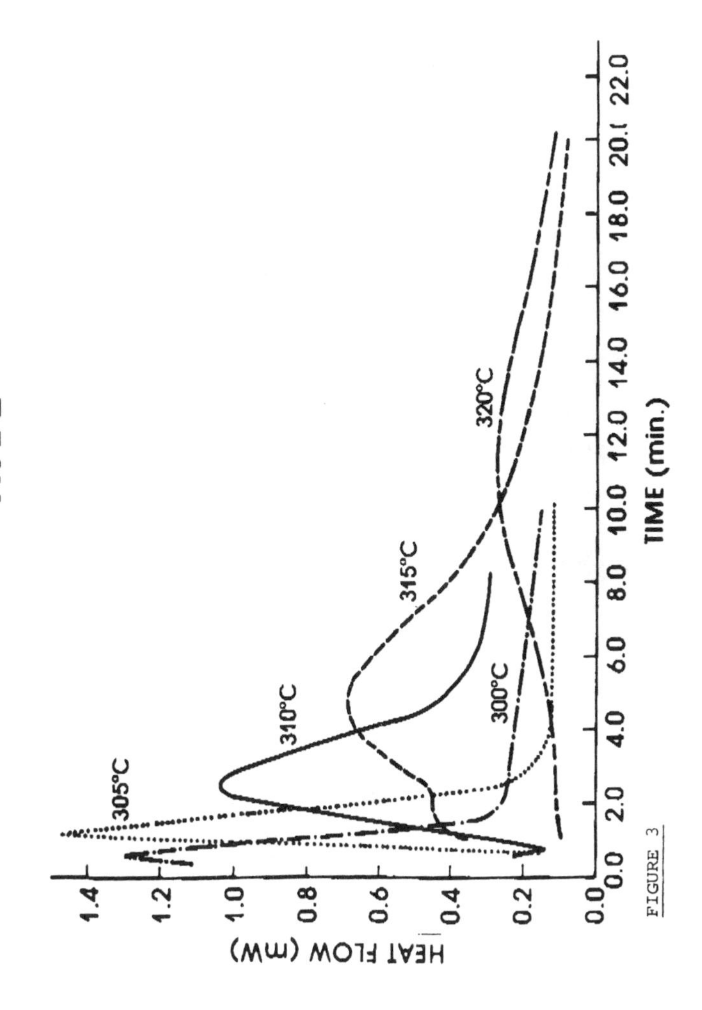
HEAT FLOW (mW)
0.0
0.2
0.4
0.6
0.8
1.0
1.2
1.4
TIME (min.)
0.0
2.0
4.0
6.0
8.0
10.0
12.0
14.0
16.0
18.0
20.0
22.0
305°C
310°C
315°C
300°C
320°C

FIGURE 3

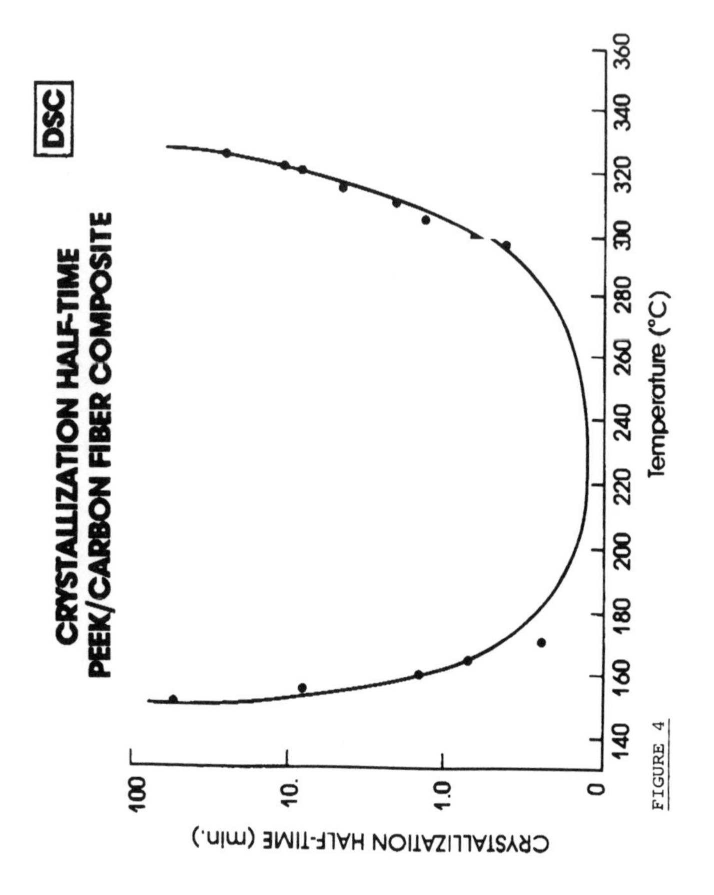
CRYSTALLIZATION HALF-TIME
PEEK/CARBON FIBER COMPOSITE
DSC
CRYSTALLIZATION HALF-TIME (min.)
100
10.
1.0
0
140
160
180
200
220
240
260
280
300
320
340
360
Temperature (°C)

FIGURE 4

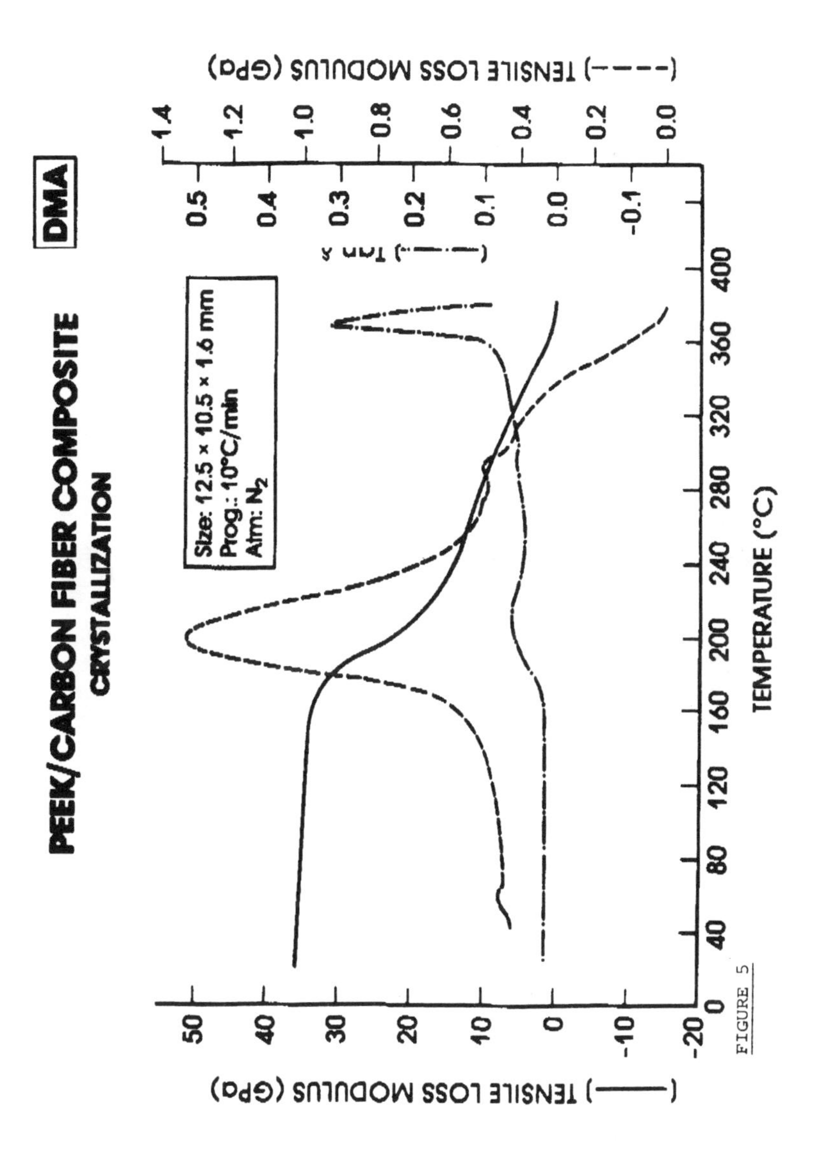
PEEK/CARBON FIBER COMPOSITE
CRYSTALLIZATION
DMA
Size: 12.5 × 10.5 × 1.6 mm
Prog.: 10°C/min
Atm: N2
(——) TENSILE LOSS MODULUS (GPa)
(– – –) TENSILE LOSS MODULUS (GPa)
(—·—) Tan δ
TEMPERATURE (°C)
50
40
30
20
10
0
-10
-20
0.5
0.4
0.3
0.2
0.1
0.0
-0.1
1.4
1.2
1.0
0.8
0.6
0.4
0.2
0.0
0
40
80
120
160
200
240
280
320
360
400

FIGURE 5

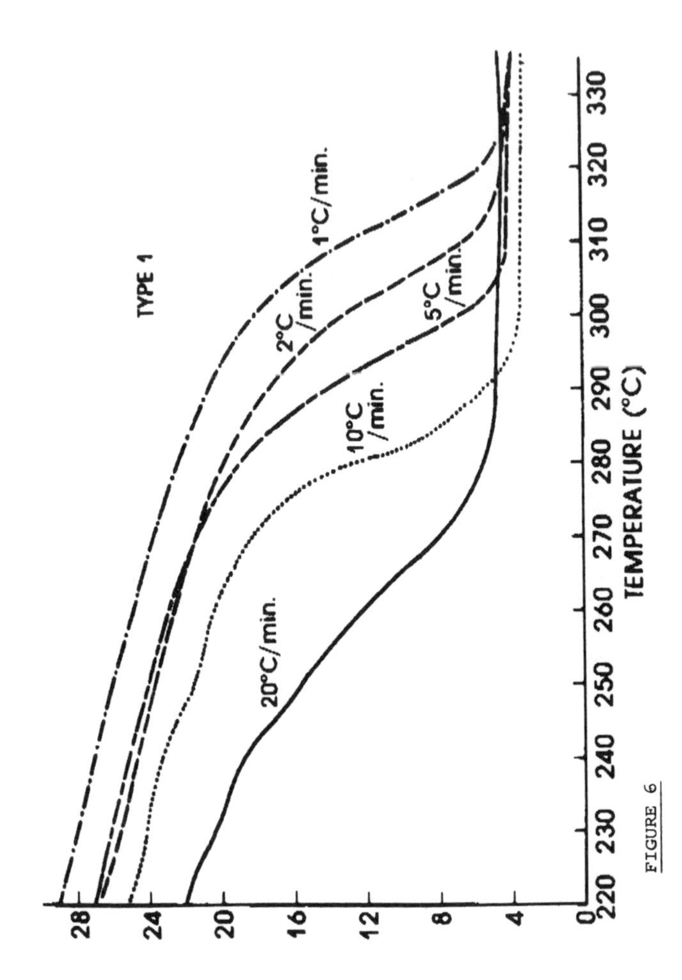
EFFECT OF COOLING RATE ON
CRYSTALLIZATION OF PEEK PREPREG
DMA
TYPE 1
1°C/min.
2°C /min.
5°C /min.
10°C /min.
20°C/min.
0
4
8
12
16
20
24
28
220 230 240 250 260 270 280 290 300 310 320 330
TEMPERATURE (°C)

FIGURE 6

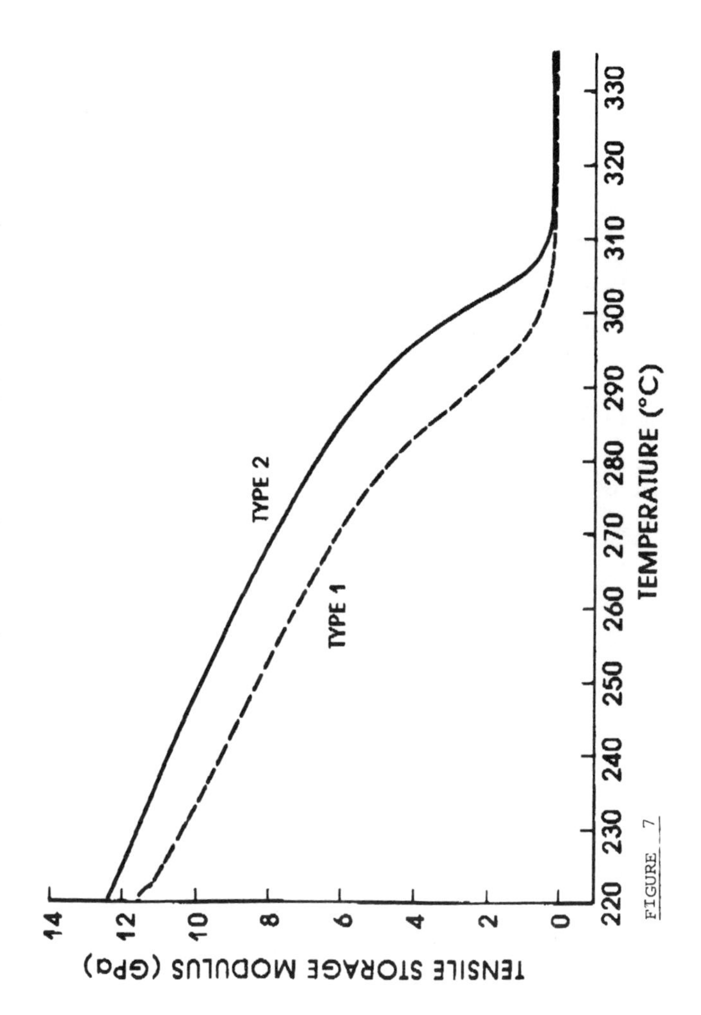
DMA
CRYSTALLIZATION OF
PEEK/CARBON COMPOSITES
TYPE 2
TYPE 1
TENSILE STORAGE MODULUS (GPa)
0
2
4
6
8
10
12
14
220
230
240
250
260
270
280
290
300
310
320
330
TEMPERATURE (°C)

FIGURE 7

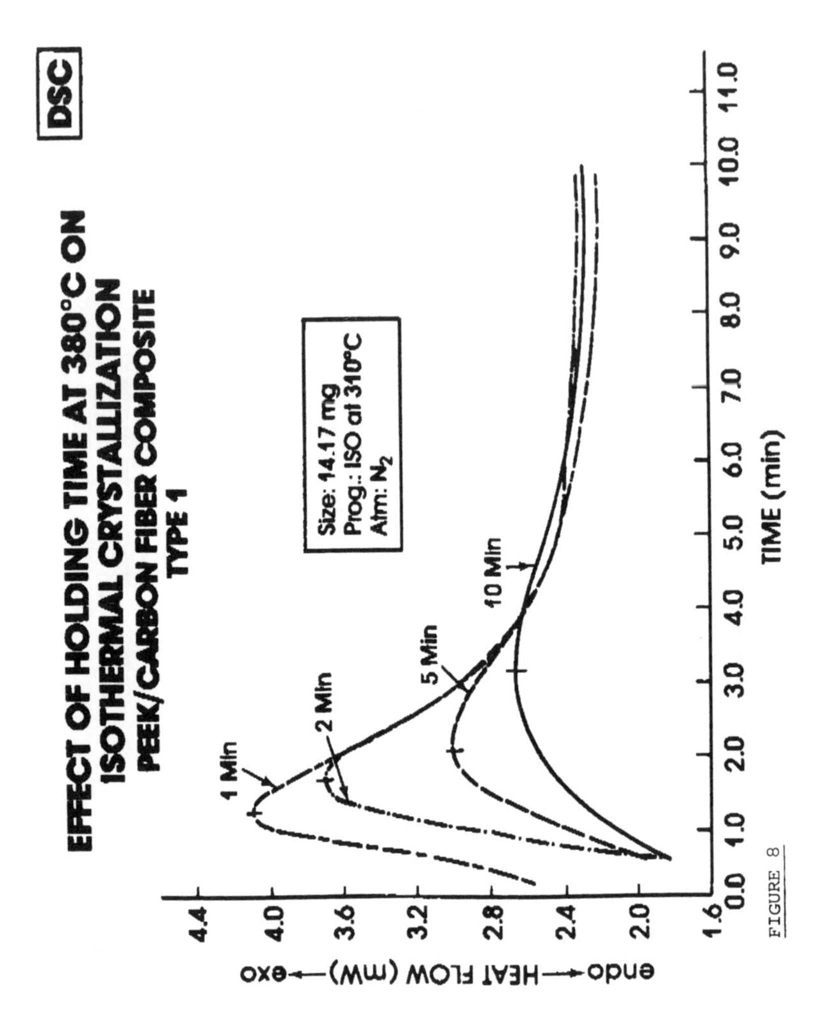

DSC
EFFECT OF HOLDING TIME AT 380°C ON
ISOTHERMAL CRYSTALLIZATION
PEEK/CARBON FIBER COMPOSITE
TYPE 1
Size: 14.17 mg
Prog.: ISO at 310°C
Atm: N2
1 Min
2 Min
5 Min
10 Min
endo ← HEAT FLOW (mW) → exo
4.4
4.0
3.6
3.2
2.8
2.4
2.0
1.6
0.0
1.0
2.0
3.0
4.0
5.0
6.0
7.0
8.0
9.0
10.0
11.0
TIME (min)

FIGURE 8

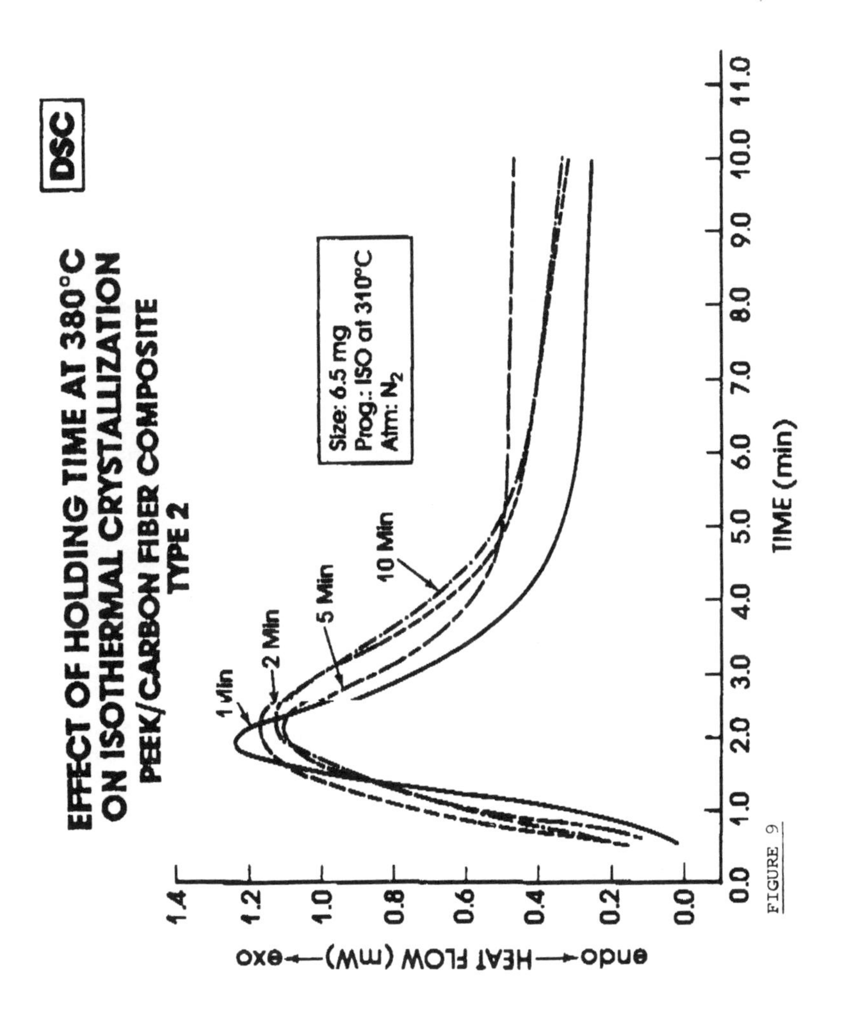
EFFECT OF HOLDING TIME AT 380°C
ON ISOTHERMAL CRYSTALLIZATION
PEEK/CARBON FIBER COMPOSITE
TYPE 2
DSC
Size: 6.5 mg
Prog.: ISO at 310°C
Atm: N_2
1 Min
2 Min
5 Min
10 Min
endo←HEAT FLOW (mW)→exo
1.4
1.2
1.0
0.8
0.6
0.4
0.2
0.0
0.0
1.0
2.0
3.0
4.0
5.0
6.0
7.0
8.0
9.0
10.0
11.0
TIME (min)

FIGURE 9

Part 3

Design and Fabrication Integration

MAJOR CFRP CIVIL AIRCRAFT COMPONENTS

(Speaker from MBB GmbH)

Paper not available at press time.

COMPONENT DEVELOPMENT FOR HELICOPTER STRUCTURES

(Speaker to be announced)

Paper not available at press time.

MATRIX SYSTEMS FOR HIGH PERFORMANCE AIRFRAME STRUCTURES

D.H. Woolstencroft, E. Heulme and J. Eastham

British Aerospace, Aircraft Group

Paper not available at press time.

PARTS INTEGRATION—ADVANTAGES AND PROBLEMS

K. B. Armstrong

Senior Airline Engineer (Stress) British Airways
London Airport, Hounslow, Middx. TW6 2JA.

The advantages and disadvantages of parts integration are considered. Examples of the size and cost of some large composite parts are given. It is concluded that the advantages are mainly with the manufacturer and the problems mainly, but not entirely with the user.

INTRODUCTION

Parts integration simply means designing to make an item from the smallest practical number of separate parts. Aircraft of metal construction are made from many separate parts bonded or rivetted together and earlier wooden aircraft were made from many separate parts glued together. Where parts are rivetted or bolted they can be separated with varying degrees of difficulty depending mainly on accessibility and on whether or not anchor nuts are used at bolt positions. It is very difficult to separate bonded or glued parts without damaging them whether or not accessibility is good.

Another method of parts integration is to make assemblies such as sandwich panels very large so as to minimise the number of joints.

For example, the one piece rudder on the Boeing 757 is 10m long and is covered by only two large carbon fibre/Nomex honeycomb sandwich panels on each side.

Yet another method is to machine metal parts from one solid billet instead of making the large part from an assembly of small ones, e.g. wing skins machined from solid billets to avoid rivetting or bonding stringers. As joints invariably add weight, parts integration is of particular interest to the aircraft industry as a method of weight saving. Again, elimination of joints results in larger individual parts.

Unfortunately, the advantages of parts integration seem to be mainly on the side of the aircraft manufacturer and the problems mainly on the side of the user. The situation is similar to that for cars and other vehicles where assembly in a particular sequence on the production line may be fairly simple. The problems lie with the user or repair man who wants to remove only the defective part, quickly and easily, without the removal of any other part. I would suggest that it should be a requirement of the design specification for any mechanical or electrical component that, "any part shall be removeable without removing any other part and without the use of special tools, insofar as this is humanly possible".

This is the repair man's dream but I hope designers will take note.

DISCUSSION

Advantages

Let us first consider the advantages of parts integration -

1. Reduced part count

   This leads to -

   less individual parts
   less storage bins
   less paper work
   less individual design drawings
   less assembly time and layout cost
   simpler assembly jigs
   less joints, so less weight

2. Weight savings of up to about 28% have been attributed to the use of composites and this is in some measure due to reduced part count in addition to the use of lighter materials.

3. Manufacturing cost is also said to be less although the very high prices of composite parts indicate that manufacturers are seeking to recoup R & D costs as quickly as possible. Where automation can be used to make composite parts, e.g. filament wound tubes, cost can be reduced. If hand layup cannot be avoided then labour cost becomes a very significant part of the total cost.

4. On those aircraft with composite fuel-containing wings, less joints mean less opportunities for fuel leaks.

5. If joints are composite to metal joints, where corrosion may occur, then less joints mean less corrosion possibilities.

Problems

1. Less parts often means larger parts.

2. Larger parts mean that damage involves -

   (a) A larger and more expensive part to replace

   (b) Higher freight charges for a larger part and larger and more expensive packing crates. Higher spare part rental charges.

   (c) If the part is rivetted or bolted then the removal of a larger part means more bolts to remove or more rivets to drill out and hence greater time delays and labour cost.

   (d) The larger and more expensive a part is, the less likely it is to be held as a spare and therefore AOG (Aircraft on Ground) action by Stores means a higher price to buy the part in a hurry and more time delays while it is in transit.

   (e) Very large parts require several people and /or cranes in order to move them or position them for assembly and the risk of handling damage is likely to be increased.

   (f) A disadvantage to the manufacturer is that larger parts require larger tooling which is costly, heavy and therefore difficult to handle and store.

   (g) Large metal billets are more prone to metallurgical defects so quality control, N.D.T. and a higher scrap rate can add to cost.

   (h) The aircraft designer, mindful of the need for maximum weight saving, sometimes tends to forget that panels and fairings have two functions. He only sees their role as providing aerodynamic cleanliness and therefore designs them as large as he can. Airline maintenance personnel only see them as large, troublesome panels to be removed to gain access for some maintenance function on a system, component or structure.

   Airlines often have to apply considerable pressure on manufacturers to get them to introduce convenient inspection/servicing access panels in some large wing/body and other fairings or cowlings to avoid frequent removal of dozens of bolts each time inspection or servicing is required. Access panels should be large enough to permit the required servicing to be carried out easily and quickly. They should also be hinged so as to trail in line of flight to avoid damage if they are not latched correctly. Latches should be of the quick release type and also rugged and reliable.

Table 1 indicates the size of some of the larger sandwich panels on Boeing 757 control surfaces.

TABLE I

Control Surface	Panel sizes (mm)	Remarks
AILERON	4,570 x 660	Full depth honeycomb construction
ELEVATOR	7,620 x 1,270	Sandwich panels
RUDDER	5,867 x 1,905 4,318 x 1,270	Sandwich panels

TABLE II

Control Surface	Cost (U.S. Dollars 1985)	Daily Rental (U.S. Dollars 1985)
AILERON	$ 65,000	$ 200
ELEVATOR	$170,000	$ 550
RUDDER	$229,000	$ 750

These prices indicate not only the desirability of repair in preference to replacement but also the need for repair methods that allow rapid return of the aircraft to service to avoid both unusable aircraft time and spare part rental charges.

Table 2 also illustrates the high price of large parts and hence one of the problems of parts integration. It also explains why spare parts are not kept on the shelf.

EFFECT OF DESIGN ON REPAIRABILITY

Regardless of whether a part is of composite construction or not there are some design aspects which make repair difficult.

Accessibility to both sides is important in repairs of all types. Even with sheet metal repairs accessibility decides whether solid rivets can be used or blind rivets become necessary.

In the case of composites access to both sides is essential if pressure or heat need to be applied to both sides. If repairs are to be carried out with hot-curing adhesives then heat and pressure must be applied to both sides. Panel removal then becomes essential. However, parts integration causes individual parts to be made larger. This means considerable numbers of bolts or rivets to be removed. Ideally, on large items like rudders and elevators the sandwich panels on both sides should be fitted with bolts screwed into anchor nuts so that removal is fairly easy. Unfortunately some composite designs have bolts without anchor nuts on one side and blind rivets on the other. It is very difficult to drill one titanium fastener from a composite panel without damaging the hole in the composite. Most large panels have several hundred fasteners! Titanium fasteners are used because they do not corrode in contact with carbon fibre. In cases like this the reduction of part count by leaving out the rivets required to attach anchor nuts is a serious dis- advantage when repair becomes necessary.

It is these panel removal difficulties which lead to a strong desire to use room-temperature curing adhesives, with or without a little heat, to repair these items. Repair can then be carried out in situ, without removing fasteners except in the immediate area of damage. Unfortunately, manufacturers are reluctant to accept room-temperature cured repairs except in lightly stressed areas.

ELEVATOR AND RUDDER

Figure 1

AILERON

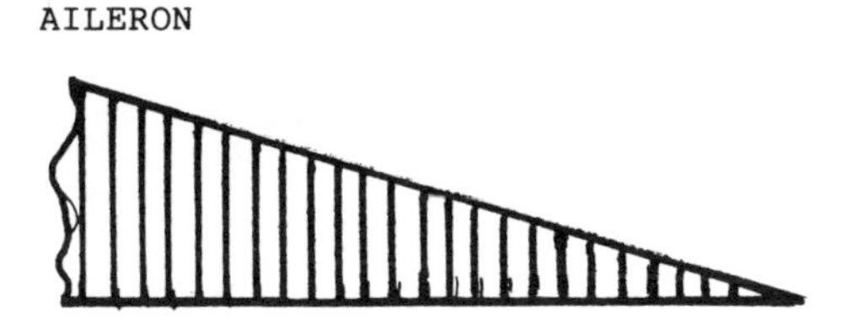

Figure 2

Development of good cold-setting adhesives and methods of warm-curing them, if necessary, may allow these repairs to be used in more highly stressed areas. This could reduce the cost of repairs and composite part ownership considerably.

The design in Fig. 1 where a separate sandwich panel is used on each side of the part is a problem if bolts with anchor nuts are not used. A real problem only exists if both skins of one panel are damaged but service experience has shown that this can sometimes be the case.

The full depth honeycomb design (Fig. 2) is much easier to repair because access to both sides is always possible. Ailerons, in particular, are not always flat and it may be necessary to make a honeycomb sandwich mould on a good part to enable the profile of the damaged part to be retained during repair.

CONCLUSIONS

1. The advantages of parts integration seem to be more on the side of the manufacturer.

2. The problems seem to be more on the side of the user.

3. The larger a part is the greater the problem of removal becomes if it needs to be repaired.

4. The larger and more expensive a part is the less likely it is to be held in stock.

   This means that in the event of damage a quick in-situ repair will be necessary.

5. Good cold-setting repair adhesives need to be developed to minimise the need for removal of large parts.

6. Damage will occur to composite parts as it does to metal equivalents and repairability should be an important design consideration.

7. Composite parts should not cost any more to repair, over their planned lifetime, than equivalent metal parts.

Keith Armstrong is a Senior Development Engineer at British Airways having joined initially as B.O.A.C in 1967. He began his experience with Vikers Armstrong in the Aircraft Division, then moved on as a Technical Officer in the RAF. His present specialisation is composite and bonded structure repairs and plastics for aircraft windows.

DEVELOPMENT AND CERTIFICATION OF AN AIRCRAFT PASSENGER SEAT

I. Tole

Futair Limited

Paper not available at press time.

CARBON FIBER COIL SPRINGS

J.C. Henry and C. Probert

National Engineering Laboratory

Paper not available at press time.

THE USE OF COMPOSITE MATERIALS IN RACING CAR DESIGN

G.P. Clarke

Advanced Composite Technology Ltd.

A brief history of how composite materials have been introduced into the modern racing car is presented. The way in which designers in many fields have made use of the increased flexibility afforded by composites is illustrated by considering the preliminary design requirements of a racing car chassis.

INTRODUCTION

In an attempt to be one step ahead of any rivals, the modern racing car has developed to be a complex machine employing many features at the forefront of advanced technology. It was therefore hardly surprising when designers looked towards composites to assist in their endless endeavour for maximized performance. It is this facet of racing car development that this paper will describe, indicating how the diverse and sometimes unique properties available with composites can be used to fulfil a designer's wildest dreams!

Figure 1 shows an exploded view of a typical Formula 1 or CART type car. The move towards turbo engines has increased the complexity of the cooling and fuel/ignition systems to an extent where weight saving on the remainder of the car is now even more important.

A BRIEF HISTORY OF COMPOSITES IN RACING

1960's Polyester resins combined with glass fibres were first used for bodywork. The ease with which complex shapes could be made by inexperienced labour was quickly recognised and when coupled with the potential for weight saving, ensured that the bucket and brush techniques were here to stay. The bodywork required little structural integrity making this low grade composite ideal.

1970's Carbon fibres and epoxy resins became generally available and rear wing end plates were manufactured as simple flat panels. The design requirements of maximum stiffness, minimum aerodynamic drag and minimum weight were ideal for a carbon fibre skinned sandwich panel construction. Typical weight savings of 60% were seen over their aluminium equivalents, a factor which was not ignored at a point of the car so far from the centre of gravity, where even small weight savings could pay dividends.

Quickly designers realised the potential for extending the end plates and making them carry the wing loads. This however changed the design requirements of the end plates as they became a primary structure. The inclusion of additional reinforcing and proper mounting inserts ensured that the basic design was sufficient for the job. Unfortunately modifications carried out on the race track in the excitement of the moment were not always within the designer's control. One such modification, which involved the re-siting of the mounting positions was not properly engineered and the resulting failure of the wing mounting assembly at 170mph was the probable cause of a fatal accident. This and similar incidents served as a warning to designers and manufacturers who were not familiar with the uniqueness of composite design and illustrated the extreme anisotropic and brittle nature of composite materials.

1977/78 Ground effect racing cars were introduced. This led to the development of the 'underwing' (an aerodynamically shaped under surface of the car) and the 'sliding skirt' which served to maintain the reduced pressure under the car. Figure 2 illustrates a typical overall aerodynamic package. This almost doubled the amount of 'bodywork' on the car and the requirement for a very stiff, lightweight, complex shaped underwing was ideal for manufacture in composites. Furthermore, as ground effect was developed, the downforces generated became greater than the weight of the car and the underwing became semi-structural.

The sliding skirts were even more ideally suited to design as a carbon fibre skinned sandwich panel. It was essential to reduce their weight and hence inertia, so they could respond rapidly to fluctuations in ride height brought about by the suspension of the car. Furthermore, in order for them to function smoothly, they had to remain flat despite the high load exerted from the pressure differences and the fact that they dragged along the track at speeds close to 200mph.

1979/80 Composite panels were first introduced into chassis. The Brabham incorporated a number of carbon fibre multi-lamina structural panels which had a nominal weight saving of 33% but were still integrated with conventional aluminium panels. Furthermore the panels were very planar although they did dispense with some of the awkward joints and additional flanges which would have been required if they had been manufactured in aluminium.

1981 to date Following on from the earlier tentative steps a chronological log becomes almost impossible due to the rapid introduction of many teams of composites in all forms.

In the 1981 season McLaren, Lotus and Alfa Romeo all introduced carbon fibre chassis each built with differing design philosophies and manufacturing techniques.

By 1984 carbon and/or aramid fibres have found their uses in all Formula 1 cars and in many Formula 2, Cart, Formula 3 and Sports cars.

Figure 1 gives a good indication of the number of parts now manufactured from composites, the most notable exceptions to date being the suspension, radiators, exhaust system, wheels and the main engine/transmission components and casings. This is not to say that developments in these areas have been ignored, but have so far been unsuccessful in giving the performance and reliability requirements demanded for competitive racing.

CHASSIS DESIGN CRITERIA

One of the most interesting and exacting components to design and manufacture in the racing car is the chassis. This part forms the structural backbone of the car and during its development has taken on many shapes and forms. The current 'bath tub' approach is seen in Figure 3 which results from the following design considerations:-

a) legislative requirements regarding the dimensions and format
b) maximum torsional stiffness
c) adequate bending stiffness
d) sophisticated aerodynamic package
e) carries the major front suspension loads - braking/cornering
f) encloses the fuel tank
g) supports the engine
h) affords driver protection
i) repairability in the field
j) no fatigue degradation

These requirements in themselves are not too exacting but coupled with additional constraints consistent with competitive racing i.e.

k) minimum weight (around 25kgs.)
l) minimum space envelope (usually limited by the size of the driver/engine)
m) subject to speeds in excess of 200mph
n) typically only 10 to 12 weeks from receipt of basic drawings before delivery required

the design is often viewed in a completely different light.

HOW COMPOSITES SATISFY THE DESIGN CRITERIA

Flexibility of shape Composites offer virtually unlimited flexibility for complex 3D structures. The only constraints arise from the timescales involved and an individual's ingenuity. The use of truly 3D structures results in the need for 3D moulds. Bucket and brush composites are often employed as a cost and time effective way of manufacturing moulds and tooling. The design of moulds and tooling is generally as important as that of the final components. Bad design at this stage can often lead to aggravation of the inherent problems of shrinkage and stability associated with this form of tooling.

The use of complex 3D shapes therefore gives the designer the freedom to tailor the aerodynamic requirements more closely to the ideal whilst giving consideration to the limitations imposed by the legislative requirements and the physical presence of a driver.

Flexibility of material properties Any complex unit, in an optimized designed form, will take on a different construction depending on the magnitude and direction of the loads at that point in the structure. The chassis is no exception, and the requirements can vary from the need to dissipate the loads exerted by an engine bolt at one point, to the mere use as a wind deflector at another. Each part of the chassis is therefore analysed with respect to the job it has to do and the materials and construction tailored accordingly. Typical construction would employ a sandwich shell with skins of 1mm thick carbon fibre laminate on aluminium or nomex honeycomb ranging from 5mm up to 30mm in thickness, depending on the particular panel in the chassis.

The ability to optimize fibre type, direction and laminate thickness at each section of the chassis gives the designer the freedom he has always yearned for and the manufacturer the complexity he could well do without.

Lightweight/high strength/high stiffness Composite materials, and in particular carbon and aramid fibres, are the obvious choice for meeting these design criteria. Extensive use of sandwich panel construction, usually using a honeycomb core of aluminium or nomex, further enhance the performance characteristics of these materials.

Impact resistance/damage tolerance The use of honeycomb stabilized panels helps to offset the brittle nature of composite panels and in some formulae its use has become mandatory in certain critical areas of the car. Aramid fibres are often used to help reduce brittle failure and their inclusion in solid bar form can give very good crash protection and reduce the tendency for composites to break apart. They however do little for the general running performance of the car, and therefore introduce a weight penalty. The designer is continually faced with the balancing of the requirements for a competitive car with those which enhance driver safety. As nobody wants to drive a non competitive racing car many of the current legislations are aimed at ensuring adequate driver safety.

The near perfect elastic behaviour of a composite structure enables damage to be localised, identified and rectified with a high degree of certainty that the remainder of the structure is undamaged and undistorted. With an all aluminium chassis crash damage at the front could distort, twist and fail the chassis at positions as remote as the rear engine mounts.

Fatigue and creep resistance It is essential that the backbone of the racing car remains rigid and strong throughout the season. Aluminium chassis suffered gradual degradation of stiffness and it was rumoured that top teams replaced a chassis after two or three races. This is not necessary for composite chassis which have been shown to retain their original rigidity after both a full season's racing and numerous repairs. Fatigue is not generally a problem as the life span of a chassis is inherently short, one or two years maximum.

Corrosion resistance The chassis experiences contact with chemicals such as brake fluids, oil and petrol during its short life. Although the resin systems need to resist any attack from these, this does not impose any severe restrictions on the design.

CONCLUSION

I have briefly shown how the unique properties available with composites have been used in the development of the modern racing car. With the particular use of the chassis, I have illustrated the initial design considerations. In order to design properly with composites, thorough consideration should be given to all the properties inherited by a particular choice of construction. Many of these properties may be used to advantage, but it is essential that those properties which are not ideal are identified and due allowance for any deficiencies made in the final design.

G P Clarke graduated with an honours degree in Applied Physics from Bradford University in 1971. He then worked for 5 years in the R&D Laboratories of Pilkington Bros. PLC., as a technologist in the stress analysis section concerned with both theoretical and experimental methods of stress analysis. His theis on 3D stress measurement in glass was awarded a PhD in Mechanical Engineering at Nottingham University. In 1979 he joined Advanced Composite Components Ltd, to assit in the rapid development of the company and in 1980 was made a director. In 1982 he became Managing Director of a newly formed subsidiary company, now Advanced Composite Technology Ltd, which specializes in the manufacture of high performance composite products.

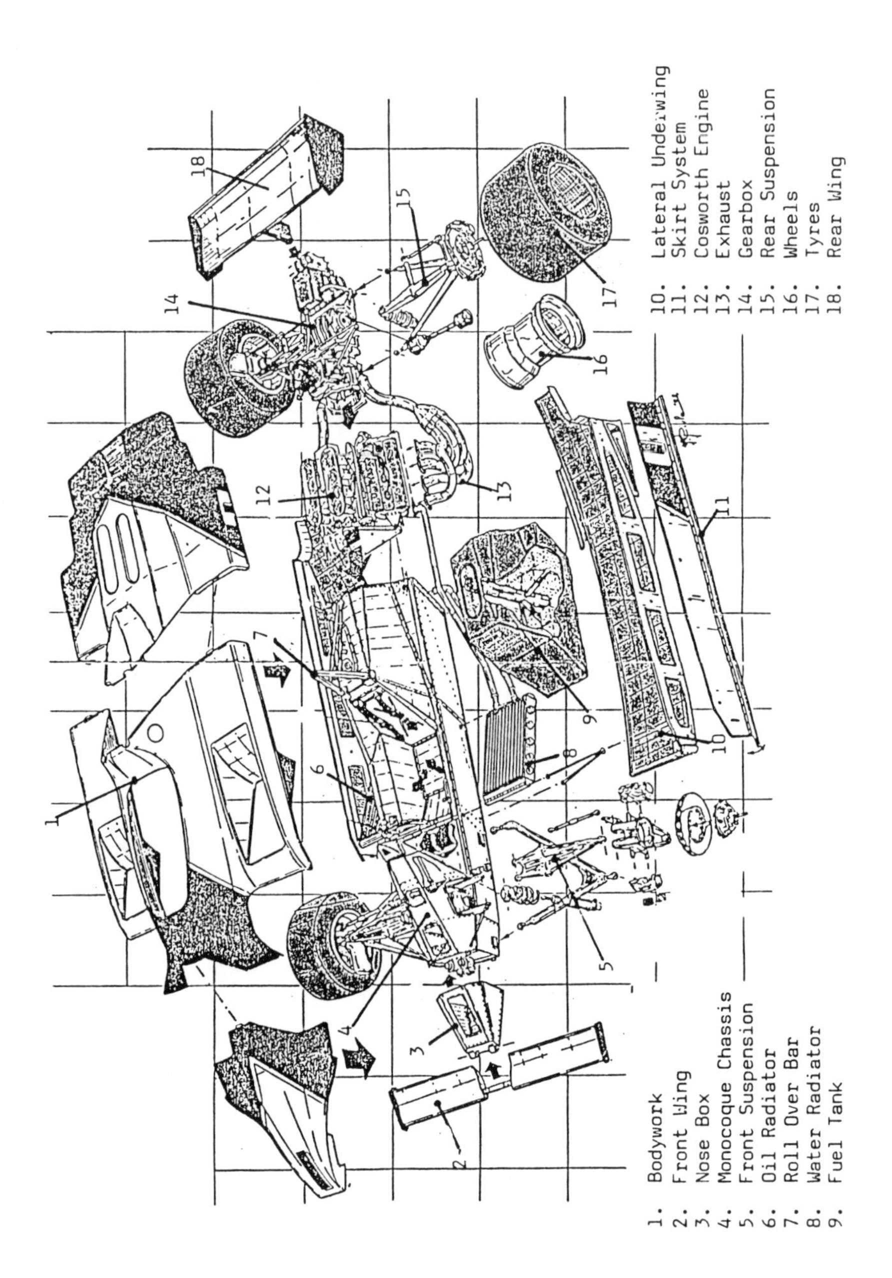

Figure 1 Exploded view of a typical racing car.

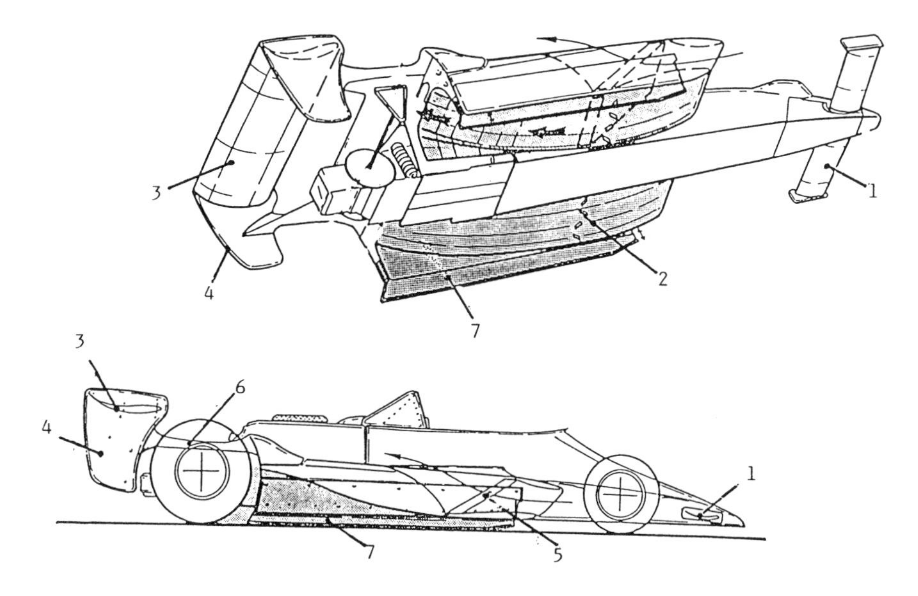

Figure 2 Typical aerodynamic package

1. Front wings
2. Venturi Section (underwing)
3. Rear wing
4. Rear wing end plate
5. Water radiator
6. Rear deck/engine cover
7. Sliding skirt

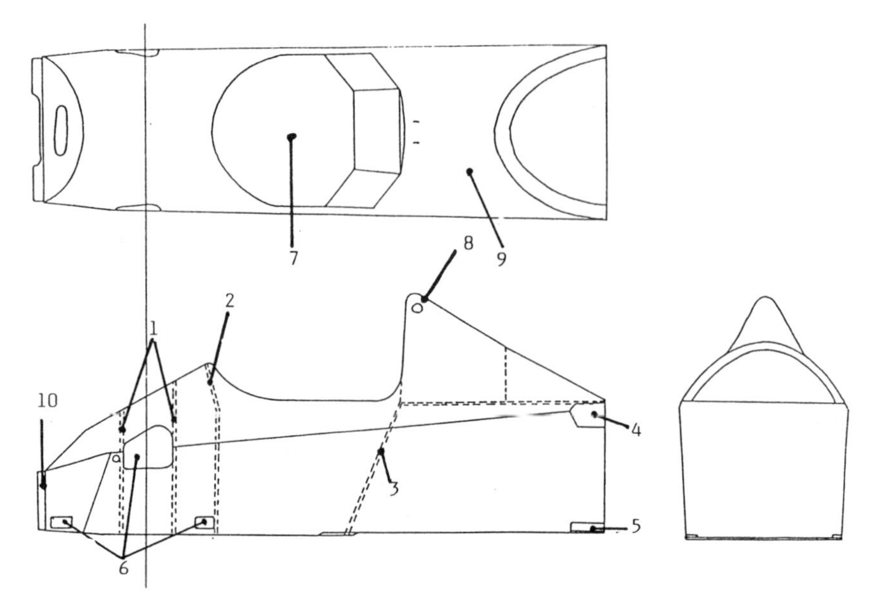

Figure 3 Schematic of a typical monocoque chassis

1. Suspension bulkheads
2. Dash bulkhead
3. Seat back
4. Top engine mount
5. Bottom engine mount
6. Suspension access holes
7. Cockpit
8. Roll over bar
9. Fuel tank
10. Front bulkhead

THE PROSPECTS FOR CARBON FIBER AN INVESTOR'S PERSPECTIVE

Sipco Huismans

Courtaulds P.L.C.

KEY AREAS FOR ANALYSIS:

- RISKS
- GROWTH PROSPECTS
- OPPORTUNITIES
- PROFITABILITY

RISKS

* YOU CAN BACK LOSERS - RR, LEARFAN

* IN A FAST MOVING TECHNOLOGY YOU CAN GET LEFT BEHIND (ESPECIALLY IN AN INDUSTRY SUBSTANTIALLY WEIGHTED TOWARDS USA DEFENCE AEROSPACE UNLESS YOU HAVE AN APPROPRIATE STRATEGY)

* IN DEFENCE ORIENTED BUSINESSES YOU ARE EXPOSED TO THE WHIMS OF GOVERNMENT

* YOUR COMPETITORS MAY BE SPONSORED - JAPAN/MITI (A KEY INDUSTRY OF THE FUTURE FOR JAPAN), US DEFENCE (ENORMOUS R & D CONTRACTS - STEALTH BOMBER, ATF, SHUTTLE ETC), FRANCE (GOVERNMENT COMMITMENT TO COMPOSITES) SEP,

* HIGH GROWTH IS ANTICIPATED AND THE PLAYERS ARE TOOLING UP FOR THE FUTURE. IF THE FORECASTS ARE OVERLY OPTIMISTIC THERE WILL BE PAINFUL OVERCAPACITY.

* AT THE FIBRE END OF THE BUSINESS IT IS HIGHLY CAPITAL INTENSIVE (£1-1.5 CAPEX PER $ SALES) AND WILL SUFFER PROLONGED CASH DRAINS.

* VERTICAL INTEGRATION ISSUES. THERE IS MUCH TALK OF THE RM MANUFACTURES MOVING FORWARD, AND SOME OF THE PRIMES MOVING BACKWARDS.

* THE JAPANESE WILL TAKE A LONG TERM VIEW - TO COMPETE SO MUST YOU.

GROWTH PROSPECTS

* THE P/E RATIO IS PRINCIPALLY AN INDICATOR OF AN INVESTORS PERCEPTION OF GROWTH PROSPECTS.

THE COMPOSITES BUSINESS IS APPROACHING BIOTECHNOLOGY LEVELS E.G.

ICI - BEATRICE	-	OVER 20
OWENS CORNING - HITCO	-	OVER 20
BASF - CELANESE	-	ARGUABLY INFINITE

BUT WHY SO HIGH?

CERTAINLY THE HEADLINE "COMPOSITES TO REPLACE METALS" IS A VERY ATTRACTIVE HEADLINE AND DEFENCE AND AEROSPACE ARE HI-TECH, HI-SPENDING, HI-GROWTH 'SEXY' INDUSTRIES.

* CURRENT FORECASTS HOVER BETWEEN 8-20% P.A. GROWTH OVER THE NEXT 10 YEARS. ALTHOUGH THIS IS LESS THAN THE LAST 5 YEARS IT IS STILL IMPRESSIVE.

SPORTS SEEN AS MATURE. AUTOMOTIVE POTENTIALLY EXPLOSIVE GROWTH BUT FORECASTERS NOT CONVINCED. DEFENCE AND AEROSPACE INCREASING SHARE WORLDWIDE FROM 60% TODAY TO 70% IN 1995.

* GROWTH ISSUES

DRIVING FORCES

- INCREASING AEROSPACE/DEFENCE ACTIVITY
- LONG TERM VIEW OF ENERGY PRICES
- COST CURVES STILL FALLING
- IMPROVING PROCESS AUTOMATION (FILAMENT PLACING, PULTRUSION, TAPE LAYING)

RESTRAINING FORCES

- CRITICAL END USES - CONSERVATISM (PERHAPS MORE ON THIS SIDE OF ATLANTIC)
- STILL A RELATIVELY NEW INDUSTRY WITH RELATIVELY SMALL DATA BASES ON MANY FIBRE RESIN COMBINATIONS.
- STILL LEARNING TO DESIGN IN COMPOSITES RATHER THAN MERELY REPLACE METAL PARTS
- LIMITED AVAILABILITY OF SKILLS (PARTICULARLY DESIGN/STRESS ANALYSIS ETC)

AND FINALLY, OF COURSE, IN THE DEFENCE AND AEROSPACE BUSINESS POLITICS ARE NEVER VERY FAR AWAY. THEY CAN TIP THE SCALES IN EITHER DIRECTION.

OPPORTUNITIES

* AEROSPACE AND DEFENCE

  - CIVIL AND MILITARY AIRCRAFT, MISSILES, SATELLITES, HELICOPTERS
  - "MEGABUCK INDUSTRIES"
  - "INTERNATIONAL" TECHNOLOGY IS VERY ATTRACTIVE TO MULTINATIONAL COMPANIES. BUT BEAR IN MIND MARKETS ARE HIGHLY NATIONALISTIC BECAUSE OF DEFENCE ELEMENT.
  - ONE CANNOT HELP BUT ANTICIPATE CONTINUED AND INCREASING ACTIVITY IN DEFENCE AND ESPECIALLY AEROSPACE.
  - THE UK AEROSPACE INDUSTRY IS DOMINATED BY BA WHICH KEEPS THE MAJORITY OF COMPOSITE MANUFACTURE IN-HOUSE. SADLY THIS DOES NOTHING TO FOSTER A COMPETITIVE COMPOSITE INDUSTRY IN THE UK. TO QUOTE DRUCKER

    "THERE IS NOTHING LIKE INTENSE COMPETITION TO FOCUS THE TECHNICAL MINDS AND SHARPEN THE MARKETING"
  - DO NOT FORGET IN YOUR STRATEGY THAT QUALIFICATIONS ARE CRITICAL.

* VEHICULAR TRANSPORTATION

  - THE GLEAM IN THE INVESTORS EYE
  - 100% PENETRATION IN FORMULA I (A COMBINATION OF AEROSPACE TECHNOLOGY, AUTOMOTIVE ENGINEERING AND SPORT)
  - WILL CF FIND A SUBSTANTIAL ROLE TO PLAY IN CARS AND COMMERCIAL VEHICLES

* SPORTS:

  - EUROPE - TENNIS RACKETS, SKI POLES, MARINE - WINDSURFING MASTS
  - F. EAST - FISHING RODS, GOLF CLUB SHEETS
  - FAIRLY MATURE - MODEST GROWTH (AROUND 4% PA)

* INDUSTRIAL

  - VARIED, BUT DISAPPOINTING

PROFITABILITY

FOR HIGH PROFITABILITY YOU WILL REQUIRE

- STAMINA TO STAY IN THE COURSE
  I.E. PATIENT MONEY

- HIGH PERFORMANCE TECHNOLOGY AT ALL
  STAGES IN THE CHAIN
  I.E. HIGH QUALITY, DIFFERENTIATED PRODUCTS.

S Huismans is a director of Courtaulds PLC. He started as a research laboratory technician with South African Iron and Steel Corporation in 1958,joining Courtaulds as shift chemist with Usutu Pulp Company in 1961 moving on through quality control to technical sales. Took a drgree in Business Administration and Economics at Sellenbosch University and came to the UK as Sales Manager, later General Manager,of the Springwood Cellulose Company in 1968. Made Managing Director of Lustre Fibres (Now Courtaulds Central Trading) late 1973 and was particularly involved in developing business in the Far East. Later also became responsible for Central Purchasing; and Carbon Fibres as a director of the Hysol Grafill joint venture company with Dexter Corporation.

SPORTS/LEISURE—MARKET OPPORTUNITIES

Risto Lapinleiumu

Exel Oy
Finland

Abstract

In the middle of the seventies Exel Oy of Finland brought a cross country ski pole of a new generation to the market. The success of the product was enormous. In a short period of time this glass or carbon fiber reinforced ski pole replaced the former metal or bamboo based ski poles in the market. With the annual sales of up to 1,5 million pairs of ski poles and a world market share of 55 % Exel is today the world market leader. And this success was especially based on carbon fiber.

The manufacturing techniques developed for ski pole tubes have since been utilized in several other summer and winter sports articles. Mass production of sporting goods has made it possible to develop such manufacturing methods, which can also be used with good results in other fields.

The main reasons that led to the use of carbon fiber in the ski pole are shortly described. Also the further development, which has partly happened in the production of new carbon fiber based products is reviewed.

The manufacturing techniques developed for the sporting goods are shortly described as well as the possibility to utilize them in industrial areas other than sports and leisure.

Background

For decades cross country ski pole has been made of nature's own fiber material, bamboo cane. Traditionally the most important qualities of a ski pole are minimum weight, adequate strength combined with maximum stiffness. Further performance features are set on impact resistance and the location of the centre of gravity.

When creating this new generation of ski poles the traditional leather and other parts made of nature's own materials were easy to compensate with engineering plastics. But the manufacturing techniques for the ski pole tube itself had to be created, and developed, by ourselves from the very beginning. The first choice of tube materials was glass fiber reinforced epoxy tube. This was because adequate strength was reached with this combination. However, the best bamboo sticks were still stiffer and slightly lighter than this man-made construction. In 1973 we made the first experiments with carbon fiber.
The table showing the growth of sales tells about the success which to large extent was achieved with carbon fiber.

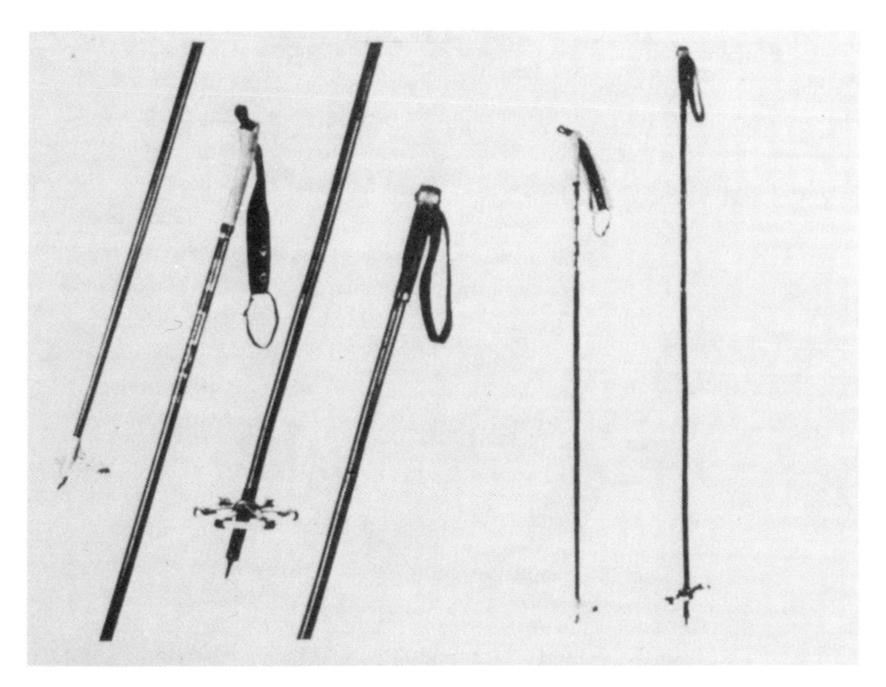

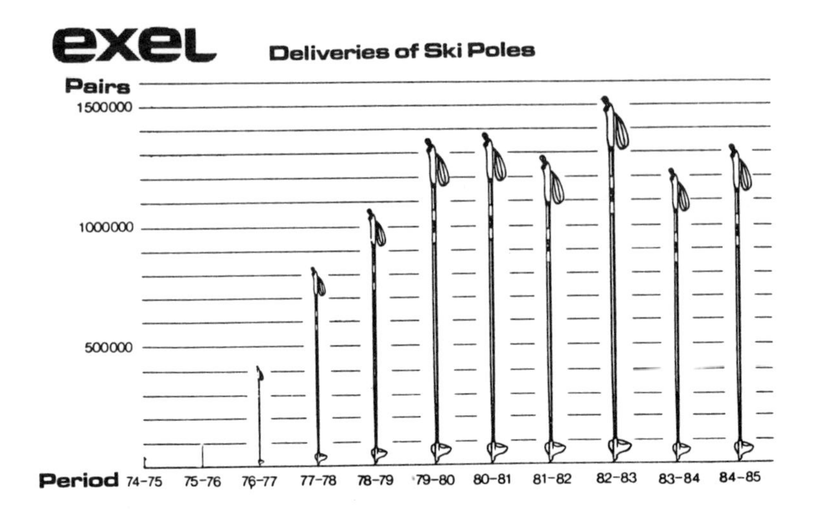

Carbon fiber has come to ski poles to stay. With carbon fiber it has been possible to develop a product that is superior when the skiers on top level are competing of meters and seconds or even parts of seconds. In theory it can be shown that when we in 1982 began to use high-strain fiber with improved qualities the weight of the tube was lowered with 7 grams per meter compared to our previous top model and simultaneously the total weight was lowered with 60 grams per pair. In theory it could be shown that this reduction in weight meant about 16 seconds per each 10 kms for the skier. So we can just imagine how well this newest evolution was accepted by skiers on the highest level. We have to remember that the cross country ski pole is a sporting goods that has to be accelerated thousands of times during a skiing period from zero speed (phase when the pole is in snow) to a speed that is about twice the speed of the skier (speed of the pole is typically 12 meters per second).

Manufacturing techniques and product features of cross country ski poles today

In few years Exel became the largest manufacturer of ski poles in the world. World market share in the western world is over 50%. The production of over one million pairs per year has led to automatization of the manufacturing processes. Today all ski poles, both cylindrical and conical, are produced with automatic, continuous production lines which are a combination of filament winding and pultrusion.

With this method, developed by Exel, the pole tube can be made of several layers. Only basic materials, carbon fiber tow and glass fiber roving, are used. These laminate layers are oriented typically to 0° and 90° degrees. A typical ski pole tube is made of 3 or 4 layers. The construction of the tube can be varied to meet the very different demands of several different user groups.

Among top skiers carbon fiber has today replaced all other tube materials from the market.

But carbon fiber brought along also some problems. When we optimized the weight, strength and stiffness relations, we came to the problem of critical impact resistance. The wall thickness of our lightest model is only 0,5 - 0,6 mm (ca. 2/100 of an inch). And this is very little for a product which can be strongly hit for example with a ski binding at the same time when the tube is under stress.

In the following a couple of solutions are mentioned, which today are used for other sports and leisure equipments as well. You naturally have to improve the matrix material and you have to optimize the laminar layers, their orientation, position in the tube etc.

But besides that even better means were required. Exel has made a glass/carbon fiber hybrid construction for skiers who can tolerate higher weight but want to have more safety in their equipments. Still this tube is lighter, stronger and stiffer than the best aluminium tube you can buy.

The other and more expensive way to solve the problem of in-adequate impact resistance it to cover pure carbon fiber tube with a Kevlar layer. With this construction the impact resistance has been doubled, but the weight remains practically the same.

Market opportunities I

The ideas and facts in the preceeding story of ski pole, its construction and manufacturing techniques have been important for Exel when creating new market opportunities. The fame of this ski pole as a superior product has given faith in carbon fiber and even demand for other carbon fiber reinforced sporting goods. According to the title of this paper I will first evaluate some market opportunities in the sports and leisure market.

For a Finnish company with such a market know-how as that of Exel, the winter sports articles have been a natural expansion area.

Glass fiber was used in skis already in the sixties, long before ski poles. Carbon fiber, however, has found its use in skis only recently in the eighties. The top models of cross country skis, in which the minimizing of the weight is essentially important, are made of carbon and/or Kevlar fiber reinforced laminates.
The weight is thus lower but above all, thanks to the stiffness of the carbon fiber, the shaping and construction of the ski is possible in a wider scale than the plain glass fiber reinforcement would allow. Kevlar fiber is however a hard competitor due to its lightness and even better vibration damping properties. However it can be used only in the bottom side of the ski because of its low compression strength.

The carbon fiber laminates used in skis can widely be applied for other purposes as well. Another big group, where laminates are used, is the ice hockey sticks and especially their shafts. The performance of wood, above all strength, is not any more sufficient in today's hard and powerful game. The stick shaft construction resembles more and more that of the skis:

sandwich-construction with wood in the middle and glass or carbon fiber reinforced epoxy laminates on sides.

Exel has also licensed a wholly synthetic shaft processing line for the biggest hockey stick manufacturer in the world, KOHO. This technique is based on the ski pole tube manufacturing process. Both glass and carbon fibers are used in this application.

Another application could be lawn hockey. But as lawn hockey is hardly at all played in Finland - we play only ice hockey - our market knowledge of this sport is small. However I believe that carbon fiber has great chances also in this sport as the material has proved its possibilities in ice hockey.

In Exel's own production range the biggest summer sports article is the windsurfing mast. Constructionally the mast is as a matter of fact an enlarged fiber reinforced ski pole in other words a conical fiber reinforced epoxy tube. The mast has been designed to stand and meet the quite different requirements of surfing sports compared to skiing. As the mast has to bend according to the sail cut and has to stand very hard blows, glass fiber has preserved its position as a leading material in the mast production.

But carbon fiber has also here brought along one interesting feature. The carbon fiber with higher stiffness than glass can be used only on the mast sides so that the mast bends in line with the sail section - i.e. in the same line as the sailboard - but the mast is vertically as stiff as possible against this line.

Thus we have made a product that simultaneously is stiff enough in the direction where stiffness is needed and suitably flexible in the direction where flexibility is needed. This quality cannot be reached with homogeneous materials like metals without a special and often expensive shaping.

If Exel now makes windsurfing masts so why not other masts as well? Yes, we have developed and manufactured prototypes of a couple of different carbon fiber masts for some enthusiasts. I'm sure there are possibilities in this field but how to realize them will be seen in the future.

Next I will introduce you to some products that are associated with water sports. One small but potential possibility are the racing oar shafts. Another and remarkably bigger group are paddles and their shafts. For both groups we have made small amounts for testing during the last year. We believe that consumption of these products will strongly increase.

Exel's other carbon fiber products with growing market opportunities are e.g. bows, in which we use the same laminates as in skis; arrows that can be made of carbon fiber by combined filament winding/pultrusion technique, and bow stabilizers which typically are conical carbon fiber reinforced epoxy tubes.

Also for sporting goods in two games with a net and a court we have delivered carbon fiber products. Badminton racket handles have been made of carbon fiber tubes and table tennis bats are strengthened with thin carbon fiber laminates.

One very large and for Exel very interesting product in the sports and leisure market product is the racing bicycle body. Carbon fiber bicycles - as is well known - have been built for racing purposes for years, but the real break-through of this product is likely to follow only if the processing technique allows to produce the frame tubes to a competitive price. I personally believe that racing cycle frames might well be a considerable new carbon fiber product group in future.

Another older and very familiar product is the golf club shaft. This product has gone through success and adversity but still you can call it a potential market opportunity.

I haven't said anything about such carbon fiber "grand products" as tennis rackets, golf club heads and fishing rods. Naturally there are even many more sports articles with hidden market opportunities.

The products that were described above are products, for which the technique originally developed for ski pole processing has created good chances. Therefore they also present real market opportunities for Exel Oy.

Market opportunities II

As I above have tried to tell, the real market opportunities and Exel's competitiveness against other materials in sports and leisure market have been realized thanks to the competitive processing methods that the company itself has developed. From the very beginning, that is early in the seventies, it has been clear that there are products which are suitable to these techniques also outside the sporting goods sector.

At least for the present the sporting goods are by far the biggest carbon fiber consumers in Exel, but growth prospects in general engineering are also very good.

As always in new technical development the first users are those, who do not consider costs so important as performance. In this group we can include military and space technology and perhaps suprisingly, private sports and leisure market. An enthusiastic sportsman does not calculate prices on the same grounds as the industry does.

How the manufacturing methods created by sporting goods can be utilized in the much wider markets depends on the following matters:

Reputation of carbon fiber products must be so good that it attracts constructors and engineers to test this material. The quality and features of a carbon fiber product correspond to the price of the product in every respect. The price comprises of the raw material price and the cost of manufacturing . Sporting goods and leisure market are paving the way for a more extensive usage of carbon fiber by increasing the raw material manufacturing capacity and on the other hand by creating more knowledge and new manufacturing processes for new applications in other fields of engineering.

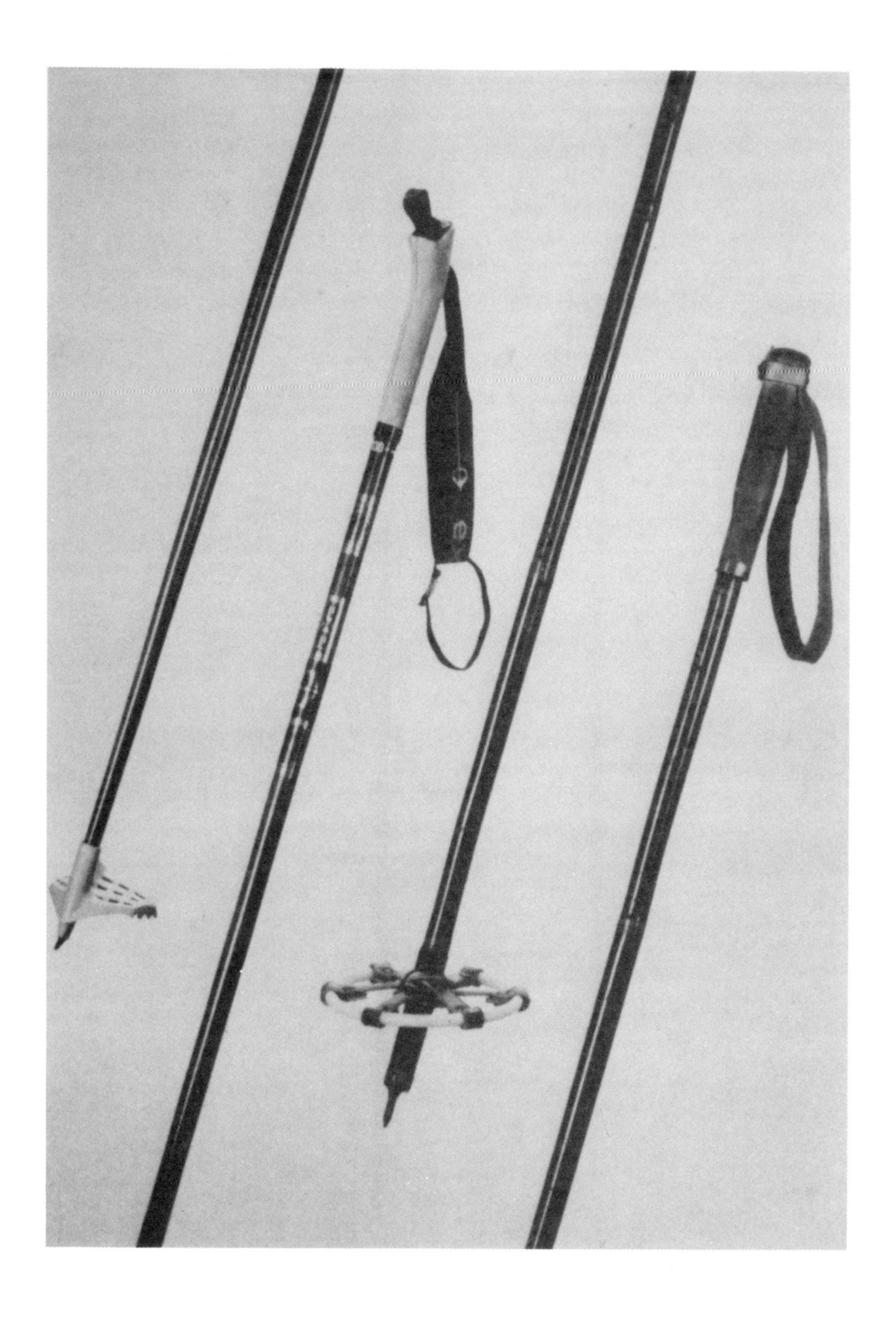

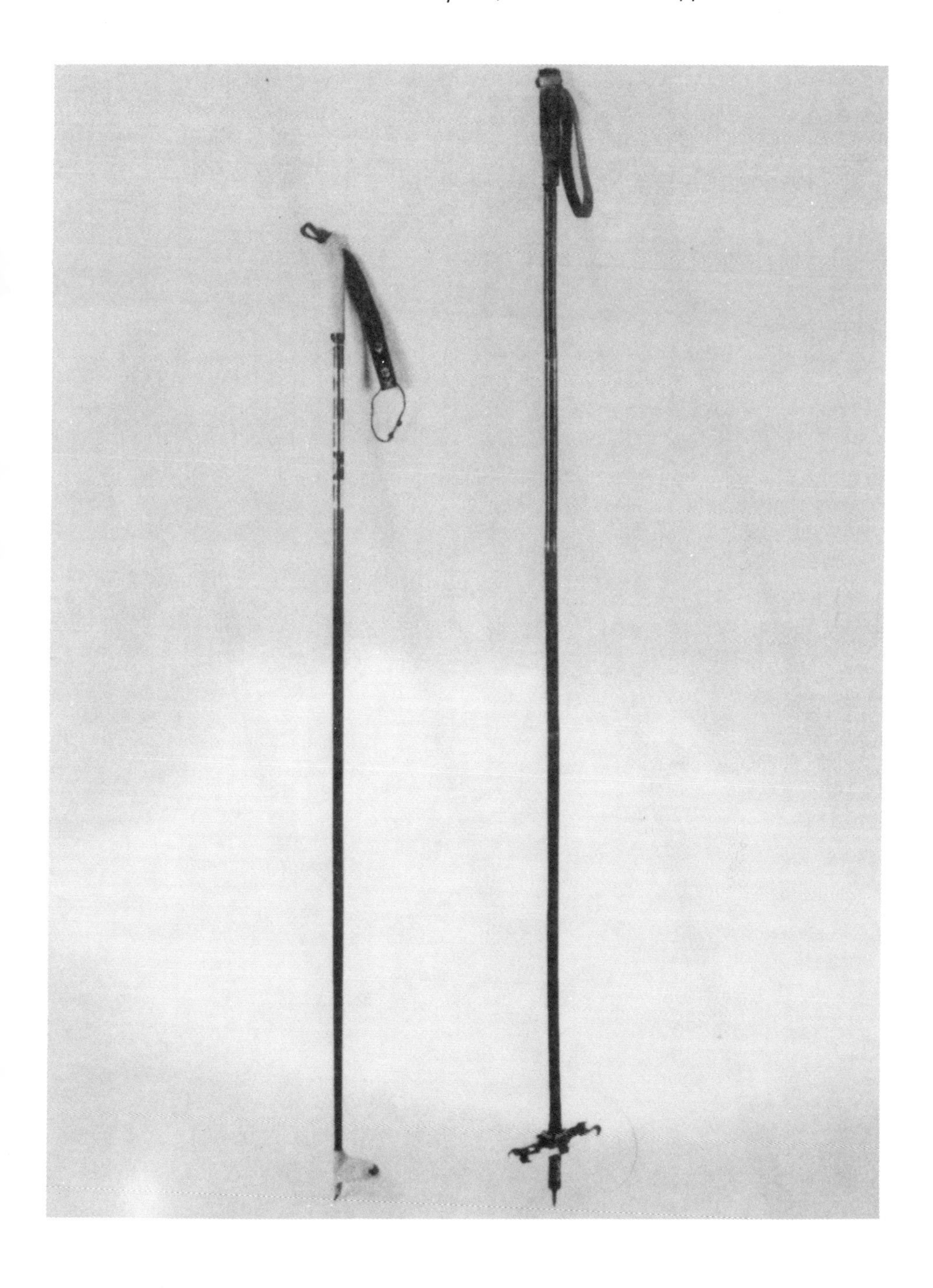

CARBON FIBER APPLICATIONS IN MARINE ENVIRONMENTS

P. Rudling

Structural Polymer Systems

Paper not available at press time.

DESIGN AND FABRICATION OF TWO SOPHISTICATED HAND LAYED-UP CARBON FIBER COMPOSITE PRODUCTS

D.W. Kent

Designability Ltd.

The characteristics and advantages to the designer of reinforced plastics are well known - to enumerate them here would be preaching to the converted. The characteristics which will be expanded upon in this paper are those involved with production over the life of a complex product, those which cause problems, and qualitatively, because of mathematical complexity, elastic tailoring of hybrid composites.

COMPOSITES IN ADVANCED AIRCRAFT STRUCTURE

The context of the paper is the work of Designability, a small Design Consultancy with the workshops in which prototypes can be built. We have neither the space nor the inclination to expand into large scale production. An efficient production organisation requires a different approach, and indeed a different mentality from that of the initial designer's and it is clearly a mistake for one engineer to attempt both functions. Initial design has to be carried out in a somewhat disorderly fashion, otherwise the same designs would be promulgated ad nauseum, but production cannot be profitably undertaken in such an environment.

The designs to be considered here are light aircraft, although the company personnel have experience of G.R.P. automobile and other work.

The two aircraft are the "Duet" and the "Leopard", illustrated in figures 1 and 2. In aeronautical terms they form opposite poles of sophistication. The "Duet", which falls into the most primitive or "microlight" category sells for £7,500, about a quarter of the cheapest metal two seat training aircraft. It is fabricated "Airfix" fashion from half shells, seat mouldings etc., glued together and mechanically interlocked. It takes off at 30 m.p.h. and cruises at 50 to 60 m.p.h.

The "Leopard" is four seat turbojet propelled, cruising at between 500 and 600 m.p.h.and forms the other end of the scale. Whilst in many ways it is a simple aircraft, the structural design is probably the most advanced in the world. This is claimed advisedly in view of its glass/carbon aeroelastically dictated design and the monolithic construction of major components. Nothing composite, repeat nothing, is rivetted together. The major components, wings, fuselage and tail are assembled into the structure by seven bolts located in integrated pick up points.

The "Duet"

This aircraft was an exercise in production design, taking full advantage of the evolutionary processes of tooling available to the G.R.P. designer. It was built from very limited private capital. The engineering requirements called for the minimum number of components, easily constructed by initially inexperienced labour. The pod shapes were sculpted from wood and rigid foam, finished and moulds taken. The product layups were all by hand, using glass cloth, tows,woven and unidirectional tapes. The strength went where it was required only. The basic shells being pared down to a few hundredths of an inch in thickness over large double curved areas to combat the aeronautical designer's continuing problem - that of weight.

Indeed, the "Duet" had this problem twofold. The normal battle with weight as an inhibitor of performance is a normally accepted part of aircraft design. The "Duet" had a further restriction in that the Civil Aviation Authority had seen fit to put a maximum empty weight limit of 330 lb. including engine on this category of aircraft. This meant that it had to carry a payload of 120% of its own weight. The best production trainer to date had carried only 76%. This was clearly an extreme test for the uses of composite structure.

Planning of "Duet" development - the evolutionary process

However thorough is one's market research, the essence of choosing to embark on a new product is the assessment of risk capital required versus market and product confidence.

Taking the most cautious approach, one builds one-off, totally by hand with no permanent jigs or fixtures. This will prove the design, produce a demonstrator for minimum outlay, but leaves a long lead time before further units of a preproduction batch can be generated. This can be an embarassing situation if the product turns out to be a good one with a ready demand.

The other extreme is to tool up with hard matched and heated moulds from square one - a very expensive and risky commitment.

The choice for the G.R.P. designer is much easier. He can build a flimsy master and take off a cheap, unsophisticated mould, build from it one or two prototypes to prove and demonstrate the design. Almost inevitably, modifications will be required, and these can be readily incorporated into the mould by cutting away the relevant area, and grafting in a new shape.

This is a very cheap process indeed, and once the design is frozen a preproduction batch can be made from this mould, also a production master made, polished and tarted up from which further full production moulds can be built. This interchange of mould and master has many advantages, one being that storage in separate buildings is an excellent insurance against fire or other damage.

As can be seen, the costs to this stage need not be high, and in keeping with confidence in the financial success of the product. The production process is still hand layup, requiring skilled operators.

The next stage is to use vacuum bagging which requires longer flanges on the moulds and a simple resin extracting system, less skill being required to obtain the correct resin to glass ratio. Further stages, each more expensive, involve the addition of heaters and later by matched moulding, the inner mould being layed up on as near an ideal hand product as possible in situ in the outer mould. Again this can often be followed by dry layups and resin injection or by prepregs for the simpler shapes.

These processes are well known, but the point being made is that the sophistication can increase hand-in-hand with the financial health of the product, and the risk to the capital involved is unusually low.

This is the basic reason, as far as "Duet" is concerned, for staying with hand layups in the early stages, in the knowledge that no unnecessary capital is at risk in the very uncertain environment in which aircraft designers work in the U.K.

The "Leopard" - the design of a sophisticated structure

Initial trial components were based on published figures for the materials used - glass, carbon and epoxy resins, with some guide on stability under load from experience of sheet metal structures.

Designability's technique is to scheme a component on the drawing board, carry out basic stressing (to engineer's theory of bending standard) then make the mould, and build and static test the component on a universal rig. The rig is basic - it consists of a 12 ft. high piece of a building-site crane jib, bolted down to a heavy ballasted concrete bed. Components are bolted to this and loaded via custom made frames by applications of several tons of small cast iron ingots.

There is little point in exhaustive stress analysis, which involves assessing fixing moments from radiused corners of the mouldings. These fixings contribute in an effective but quite indeterminate way to the stability of the component under load. Given a mould which can produce components of varying layups, and capable of modification, it is much cheaper and often quicker to make a component and test load on the rig. Depending on the complexity, function and numbers to be made is the decision to continue the test to destruction or to merely over-proof load, then use if no damage has occurred.

The first major and unique load path components designed in this way were the spars of the "Duet" aircraft. We broke almost a dozen, including one disastrous carbon-only specimen before settling on an acceptable layup.

Tension characteristics of carbon and glass were found to be well up to quoted figures, but compression stability was well below that forecast. It should be noted that Designability is very cost conscious and large chunks of expensive carbon fibre were not considered economically acceptable unless absolutely essential for strength requirements. In the event, we found that the weakness of compression components was in the resin and reinforcement by carbon had little advantage over that of glass except where it could be used to augment stability.

Another problem with carbon is that its stiffness inhibits even distribution of load across a section from a discrete load point or points. This loads the resin unduly in local areas causing early failures. The lower modulus of glass renders even distribution to occur much more freely.

Another practical difficulty we found with carbon was that the finer close packed strands were much more difficult to wet through with resin - and resin often had to be warmed, thus reducing its pot life.

The best sections for compressive loads, from cost and structural effectiveness view points turned out to be basically glass, but"tuned" with a very small percentage of carbon, placed to inhibit buckling. The glass can thus be "beefed" up in line with the prime load patterns - even to the extent of diverting the load paths into the most acceptable areas of the component.

The "Leopard", the next project to be undertaken is a relatively complex and expensive structure, and the cost of each component made necessary a careful analysis of the value of destructive, rather than proof testing.

The structure was stress-analysed in much more detail, using the stresses obtained during "Duet" testing as a data base. First a "representative component", a tailplane was designed purposely on the light side and tested - to destruction. In the event several were built, trying out various layups in the load-carrying skins and spar webs. It was discovered that a minimal amount of strategically placed carbon rendered the skins stable and this, and other lessons were then applied to the design of the remainder of the aircraft. The other large components - the fin, wings and fuselage were "proof tested" only, then built up as the flying prototype.

The fuselage on this first aircraft was overdesigned as the critical loads were from pressurisation, rather than flight loads, and the prototype is not to be subjected to internal pressure.

The "Leopard" wing - elastic tailoring

The wing design was a very interesting exercise, as using conventional techniques, the structure would have been certain to flutter at the higher end of the aircraft speed range. The only practical position of the spar was about 65% of the wing chord back from its leading edge and the air loads centre at around 25% of the chord. The loads then twist the wing nose up about the rear spar, increasing the wing's angle of attack to the airflow, generating more lift, more twist, until the flow breaks down, whereupon the wing flips back and past its original position to build up the loads in the opposite sense, an often very rapid and destrúctive oscillation resulting.

It was necessary to move the flexural axis forward, near to the centre of effectiveness of the air loads, and this was done by using carbon cloths and unidirectional tapes in opposition to the basic glass structure to give a "virtual flexural axis" a long way forward of the main aft-located spar.

Various techniques were tried, including a one-way torsionally stiff wing box which had spirally wound tapes over its outer surface. This did not work, so the angle of the tapes on one side was reversed to give the same planform sense to the angles top and bottom. This did not work either, so we had to generate preferential load paths over parts of the wing which tended to transfer the loads in the outboard part of the wing to a suitably graded false front spar, and also suppressed the outboard half of main rear spar.

This is rather involved, but demonstrates how modulus of carbon versus glass can be used to tailor the stiffness, and so the load paths and deflection patterns in a structure. In aircraft, this is of special importance in view of the weight problems with high resultant stresses and deflections generating those of flutter. An aircraft can dismantle itself in a matter of two or three seconds if the divergent process described does occur.

The technology is still in its formative stage, even so the mathematics are of a complexity such that powerful computer analysis is necessary to the finding of solutions. It has been said that of every new fighter design of the last twenty years, one aircraft has disintegrated during development flying, probably due to rapid flutter onset.

Thus the design of high speed aircraft can be a daunting prospect, but the increasingly understanding of aero-elastics, together with the stiffness tailoring possible with hybrid composite construction as applied to the "Leopard" does offer the tools for coping with this problem.

Composites for landing gear

Another use of carbon stiffened glass components is in spring beams for aircraft landing gear. It was shown many years ago by Cessna of America that a leaf-spring with no damping component could be used for absorbing the vertical kinetic energy of a landing aircraft. The damping comes effectively from the aircraft itself - the large wing areas and its dynamic behaviour on touchdown. Metal cantilever springs used by Cessna are exceedingly heavy, even if very high tensile steel is used.

Glass leaf springs have been used for some years, first on the Swedish Vipan light aircraft which used half of a 360° filament wound component of some 8 ft. diameter. Wheels were bolted to the ends of each half of the semi-circular member which was then clamped to the belly of the aircraft.

Glass leaf springs have to be thicker and produce more aerodynamic drag than their metal counterparts and are still heavy but when stiffened with carbon, can be made thinner and lighter.

Without this technique for design of the traditionally heavy landing gear, the "Duet" could not have been made anywhere near the 330 lb. limit imposed by the Aviation Authorities.

The layup is by hand, the laminations being placed in a profiled trough. First cloth laminations are applied, some at zero degrees to the axis of the spring, some at 45° to give torsional stiffness. The bending material, unidirectional carbon laminates, then the bulk of the glass is packed in and finally the free edges of the initial cloth laminates wrapped over. A cap is then clamped over the trough to give a clean finish to the upper surface of the component.

These springs are incredibly strong and take a heavy pounding during training operations where rough landings and taxiing take up a considerable proportion of the operating life of the aircraft.

The nose gear spring of the "Duet" posed an interesting problem as it had to cope with heavy torsions as well as bouncing loads during ground manoevring. The spring which curves over the front of the wheel, then bifurcates either side to form integral trailing links distorted badly when sideloads were applied during turning. This again was stiffened in torsion by the use of carbon strategically placed so that the energy absorbing flexure was not inhibited. Some more work needs to be done on this component, but the effective techniques are clearly established.

Further design work

The message then, of the latter part of this paper, is that carefully designed hybrid monolithic structures can be made to carry out the functions of mechanisms normally needing several components bolted together. As with many developing techniques, the aircraft industry is being one of the early exponents of the methods of design which will find application in many other spheres (note that the J.C.B. earthmover is based on hydraulic controls which were developed during World War II for aircraft undercarriage and control actuation).

Many structures which are required to accept distortion or absorb energy can be designed tc do so within their own material, rather than be bolted via springs or rubber absorption units - and do so only in senses or directions which do not impair their basic functions.

Sports equipment is an obvious example. Another is a bicycle frame which would absorb vertical energy to soften the ride without deflection out of its vertical plane which wastes the rider's energy input.

Aside from flexible products to components where lightness and cost are essential factors, carbon stiffened glass can lead to very elegant and effective designs. It could be that the use of small quantities of relatively expensive carbon could pay for itself in reduction of bulk of the component, thus in the cost savings of the reduced resin requirement.

There is certainly no shortage of scope for the ingenious designer.

DUET

Mockup of Chichester-Miles Leopard business jet (M. W. Kent)

LEOPARD

FIGURE 1

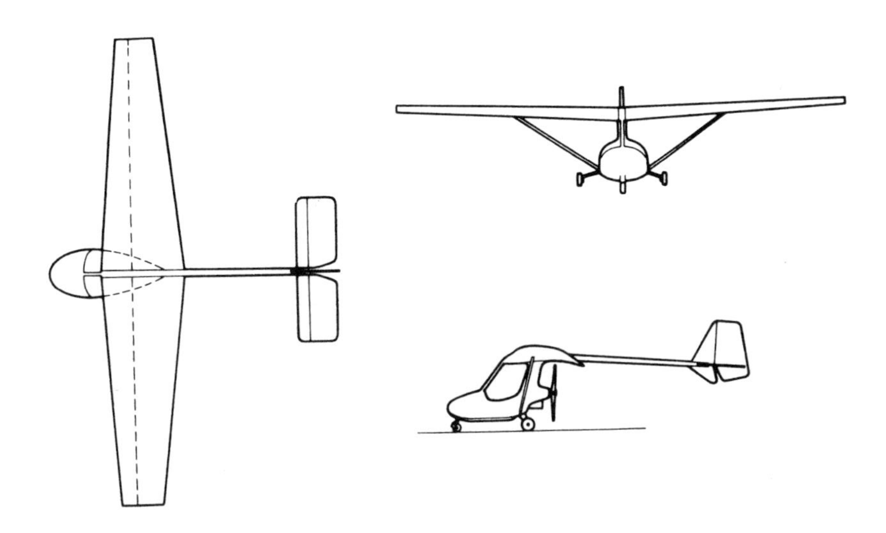

Proposed production specification

Span	36 ft (Wing Area 162 sq.ft.)
Length	22 ft
Empty weight	330 lb
Payload	385 lb
Fuel capacity	5 gallons
Maximum all-up weight	750 lb
Maximum wing loading	4.6 lb/sq.ft.
Load factors	Design + 4g – 2g Ultimate + 6g – 3g
Stalling speed	31 mph at max. AUW. 27 mph solo
Maximum rough air speed	60 mph
Maximum speed VNE	90 mph

Measured performance figures on prototype DUET G-MBWH

These figures, corrected to ISA are derived from values measured during flight tests by simple direct and square root ratios.

The airspeed indicator was calibrated with a trailing static point over the speed range required. Several runs were carried out with different pilots to obtain these results. Tests were conducted at an all-up weight of 660 lb often in weather conditions which were anything but ideal.

The engine used was a 430D Cuyuna producing 32 hp at 6,000 rpm driving a 2-blade wooden propellor of 45.5 in. diameter and 15° pitch through a 2:1 reduction ratio. Propellor maximum tip speed is 595 ft/sec.

	MEASURED ON PROTOTYPE 145 sq.ft. wing giving 4.5 lb/sq.ft. at AUW	PRODUCTION AIRCRAFT (ESTIMATE) 4.6 lb/sq.ft. wing with 40 hp Rotax engine
STALL SPEEDS (TRUE)		
At design AUW	31 mph	30 mph
1 occupant	27 mph	26 mph
RATE OF CLIMB		
At design AUW	312 ft/min.	580 ft./min.
1 occupant	387 ft/min.	750 ft/min.
TAKE OFF DISTANCES		
Tarmac at design AUW	190 ft.	150 ft.
1 occupant	155 ft.	120 ft.
Dry grass at design AUW	320 ft.	250 ft.
1 occupant	250 ft.	200 ft.

Measured full throttle level flight speed at design AUW was 58 mph.

FIGURE 2 EARLY DUET SPECIFICATION

INTERLEAF TECHNOLOGY FOR CARBON COMPOSITE AIRCRAFT PRIMARY STRUCTURE

R.B. Krieger, Jr.

American Cyanamid Company

Carbon fibre composites can save weight over aluminum in aircraft structure. This savings could be magnified if the composite were "tougher", i.e., more tolerant of damage and stress concentration. To accomplish this, the Interleaf Concept has been developed. A thin, high elongation, resin interleaf is placed between each fibre ply to provide toughness. Each ply has its fibres in an epoxy matrix for high hot-wet performance. The two resins remain discrete during cure and so interleaf toughness does not compromise matrix hot-wet compression strength.

INTRODUCTION

The use of graphite composites in primary aircraft structures has not reached full potential. Current structure in production is limited in ultimate design strain (strength) allowables because of toughness deficiency. The ultimate strain allowable for current production is .004 inches per inch. This value appears insufficient for interesting weight saving. If quasi-isotropic tension modulus is taken as 8,000,000 psi, the ultimate tension stress becomes 8,000,000. x .004 equals 32,000 psi. This is half that of 2024-T3 aluminum. Since the density of graphite composite is no better than half that of aluminum, weight saving does not appear convincing. Graphite can save weight by fibre orientation, and also in secondary structure, where the heavier aluminum is not used to its full strength. For full potential, we must extend the allowable strain from .004 to .006 or better. This demands new levels of toughness without the usual sacrifice of hot-wet compression strength. Modified epoxy resins have reached proper toughness levels, but to no avail, because of inadequate hot-wet compression. The Interleaf Technology <u>does</u> have decisive hot-wet capability. It is extremely tough; the data in this paper shows it is possible to reach ultimate strain levels between .006 and .008 inches per inch.

THE INTERLEAF CONCEPT

Webster defines toughness as "having the quality of flexibility without brittleness; yielding to force without breaking." The second phrase is, perhaps, the classic definition of toughness.

If we consider a tough aluminum alloy, say 2024-T3, the "yield force" occurs at two thirds of ultimate strength. Prior to this, the stiffness curve is linear with a strain at the yield point of 0.4%. A sharp drop in stiffness occurs at the yield point. The curve continues, (essentially linear), to failure, at an ultimate strain of 7% to 12%. It is this huge increase in strain with no drop in strength that constitutes toughness. By contrast, the carbon fibre stiffness curve is linear to failure. There is no yield. Ultimate elongation barely exceeds 1.5%, with 2% as the inherent ceiling. The fibre cannot be considered to have toughness, now or in its future development. It follows that laminate toughness must come from the matrix. Epoxy chemistry is defeated by hot-wet compression weakness. To discover the next step, a study was made of impact failure modes in the best epoxy formulations. Fig. 4 is a careful section drawing of a panel impacted by the popular energy of 1500 inch pounds per inch of panel thickness. Site 1 is tension failure of surface plies. Sites 2, 3 and 4 are all interlaminar shear failures, ("splitting"). Site 2 failure is caused by the forces released by 1) tension failure. Site 4 is at the neutral axis (central plane) where shear stress due to bending is a maximum. Site 3 is secondary, or even tertiary, to the others. These interlaminar shear failure are predominate at all impact levels, including the "punch-out" mode and through-penetration damage. The remedy or solution suggested by all this is shown in Fig. 5, the description of CYCOM®HST-7. The essence of the material is a tough resin interleaf between each ply. This is accomplished by adding a thin sheet of tough resin to one face of the impregnated fibre tape. This high strain interleaf provides toughness. The resin on the fibres themselves provides the hot-wet compression strength. The interleaf is formulated so that it remains discrete, well bonded, void free layer after it is co-cured with the fibre matrix. Fig. 6 presents photo enlargements showing the discrete interleaf compared to a conventional laminate.

TEST METHODS

There are two decisive tests, 1) Compression After Impact and 2) Hot-Wet Compression. Quasi-Isotropic laminates are used for both tests. Quasi-Isotropic is a fair comparison to isotropic aluminum. Zero degree fibre orientation is the compression test most searching for the matrix resin. However, real structure is seldom 0° oriented. Weight saving by fibre orientation, is some degree, is a valid concept, but Quasi-Isotropic is selected as a properly conservative basis for comparison. Fig. 1 shows a very compelling reason for selecting Quasi-Isotropic over 0° fibre orientation. It can be seen that large differences in 0° fibre tests are not reflected in Quasi-Isotropic tests, and so will not appear in real life structure. Fig. 2 shows the device used to produce impact damage on the test panel. The panel is 4 inches x 6 inches and is laid up as (+45/90/-45/0) 3S with 190 grade tape and Hercules AS6 fibre. After impact, the panel is tested in compression, with simple edge supports. Fig. 3 shows the hot-wet compression specimen. The test is run at 200°F after specimen immersion in water at 160°F for 20 days. This gives a water pick-up of approximately 1.5%. The laminate ply arrangement is (+45/90/-45/0)S.

TEST RESULTS

The decisive tests are compression after impact and hot-wet compression. Fig. 7 presents data for these tests for several versions in the development of CYCOM®HST-7. In order of ascending values for hot-wet compression, these versions are: 1) Grade 145 tape at 47.5% resin content, 2) Grade 190 tape at 45.1% resin content, and 3) Grade 190 tape at 41% resin content. As expected, the lower resin content gives the highest stress. Much statistical work has been done to establish the minimum weight for the interleaf. Standard production material for CYCOM®HST-7 will have a nominal resin content of 41%.

WEIGHT SAVING OF CYCOM®HST-7 OVER 2024-T3 ALUMINUM

The pertinent properties for aluminum are taken as .098 lbs. per cu. in. for density and 64,000 psi ultimate allowable for compression. For CYCOM®HST-7, the pertinent data is 41% resin content, and, for the critical ultimate strength, 73,000 psi compression at 200°F and wet. First, we must calculate the density of the composite. We use this formula:

$$\frac{285}{(W_r + 2.59) \times 1728} = .056 \text{ lbs/cu. in.}$$

where W_r = .41 = resin weight fraction.

Next, we calculate the weight saved as

$$\left[\frac{\left(.098 - \frac{64000}{73000}\right).056}{.098}\right] \times 100 = 50\%$$

where the parameters for density and stress allowables are apparent.

INTERLEAF STRAIN ACCOMODATION

The presence of a high elongation (tough) interleaf sharply improves a laminate's performance beyond the foregoing toughness concepts. The interleaf represents a finite thickness of resin which is perhaps twenty times thicker than the resin between plies in a conventional laminate. This means the interleaf is far less stiff and will eliminate stress concentrations which otherwise would produce premature matrix failure. Fig. 8 describes two of these situations in detail. Ref. 2 provides the theory for shear distribution.

CONCLUSIONS

1) Conventional composites (having a single matrix system) are inherently limited in toughness. The constraint of hot-wet performance produces a low strain matrix. The low resin content produces thin (stiff) resin bonds between fibres and between plies. These thin, stiff, bonds are Achilles Heels. 2) The interleaf technology eliminates the stiff matrix problems. CYCOM®HST-7 has twice the toughness of current production graphite composites. Furthermore, it maintains a 50% weight advantage over aluminum in the critical allowable of hot-wet compression.

REFERENCE

1 "Stress Analysis of Metal to Metal Bonds In Hostile Environment" Adhesives Age Magazine, June 1948, Vol 21, No. 6.

BIOGRAPHY

Raymond B. Krieger, Jr., graduated M.I.T. with a B.S. in Aeronautical Engineering in 1941, and from then until 1953 was with the Glenn L. Martin Company. Until 1956 he was with Luria-Cournand as Chief Engineer. From 1956 to the present he has been with American Cyanamid Company, Engineered Materials Department, first as Chief Engineer, then Sales Manager, now as Technical Manager. His duties encompass criteria development and testing of structural adhesives, technical service to the airframe industry including design, processing and quality control aspects of structural bonding, metal to metal, sandwich and structural plastics. He is the author of numerous papers on structural bonding and composites.

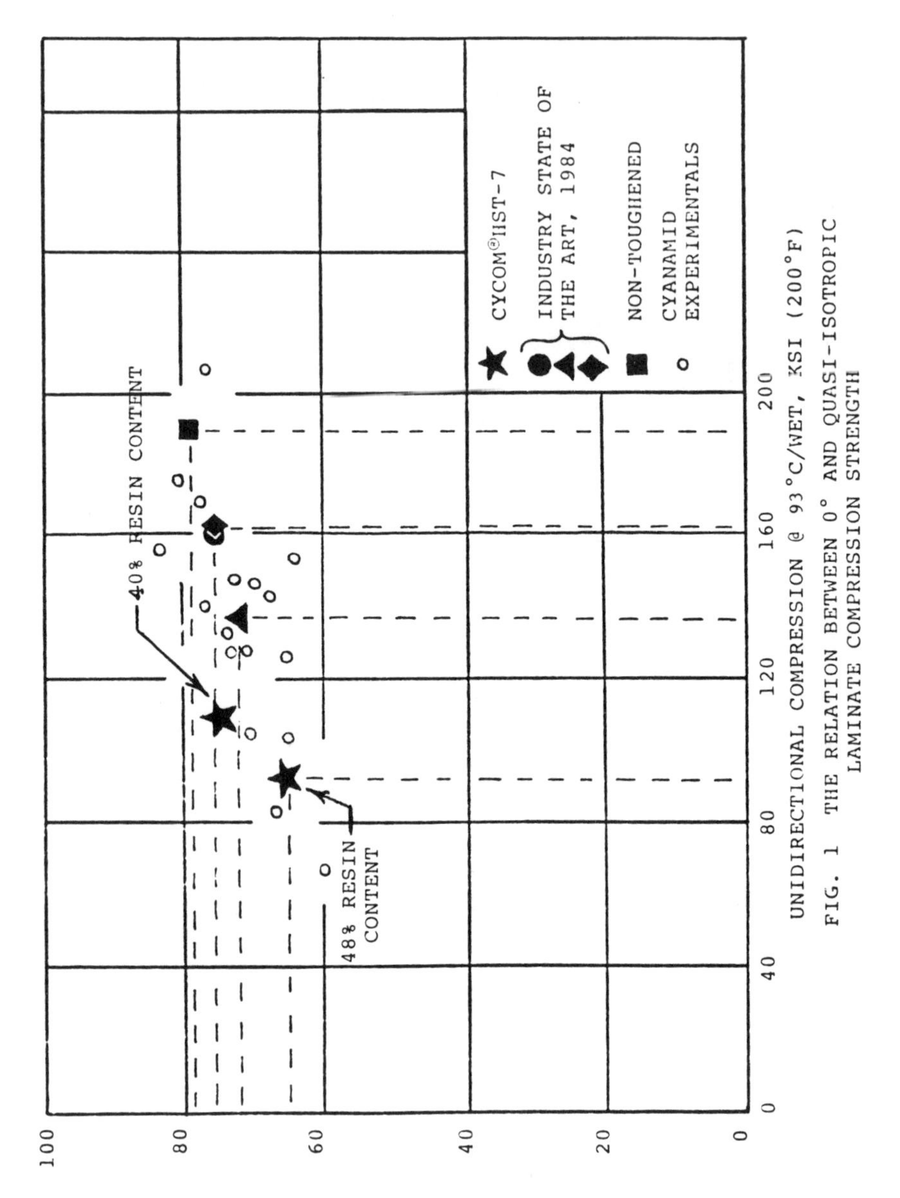

FIG. 1 THE RELATION BETWEEN 0° AND QUASI-ISOTROPIC LAMINATE COMPRESSION STRENGTH

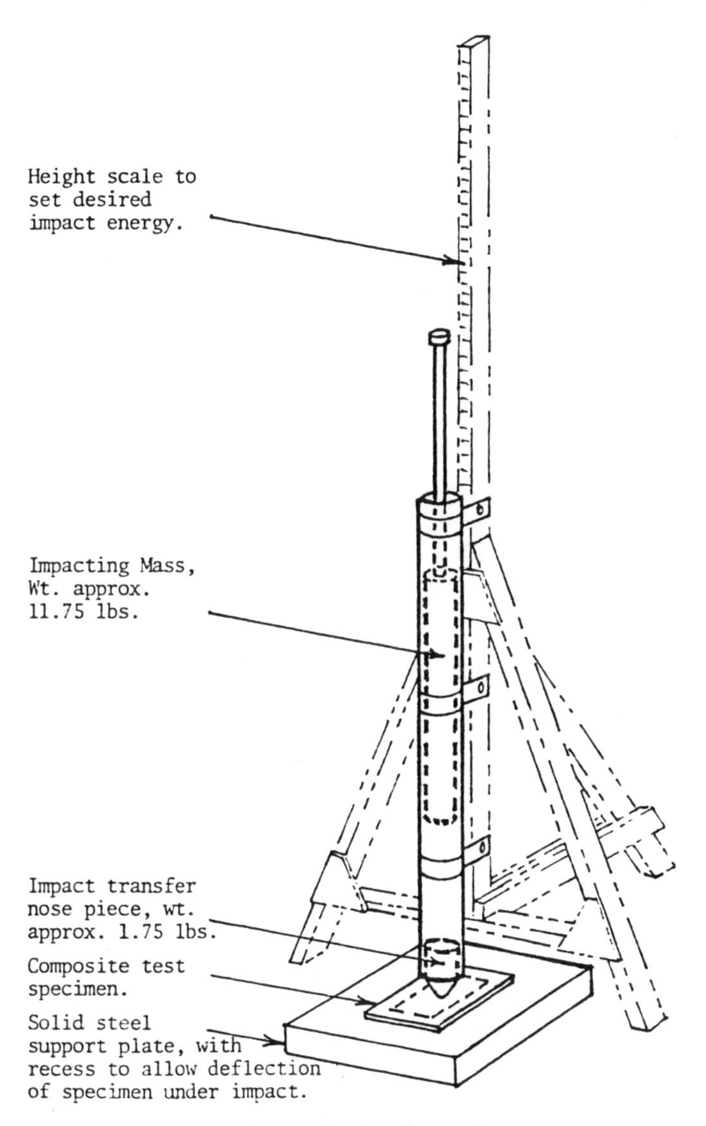

FIG. 2 - Fixture for Producing Impact Damage on Composite Compression Test Panels.

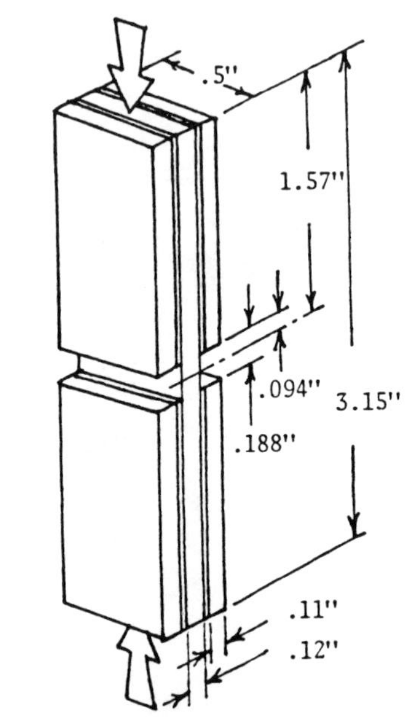

FIG. 3 - Hot-Wet Compression Test Specimen.

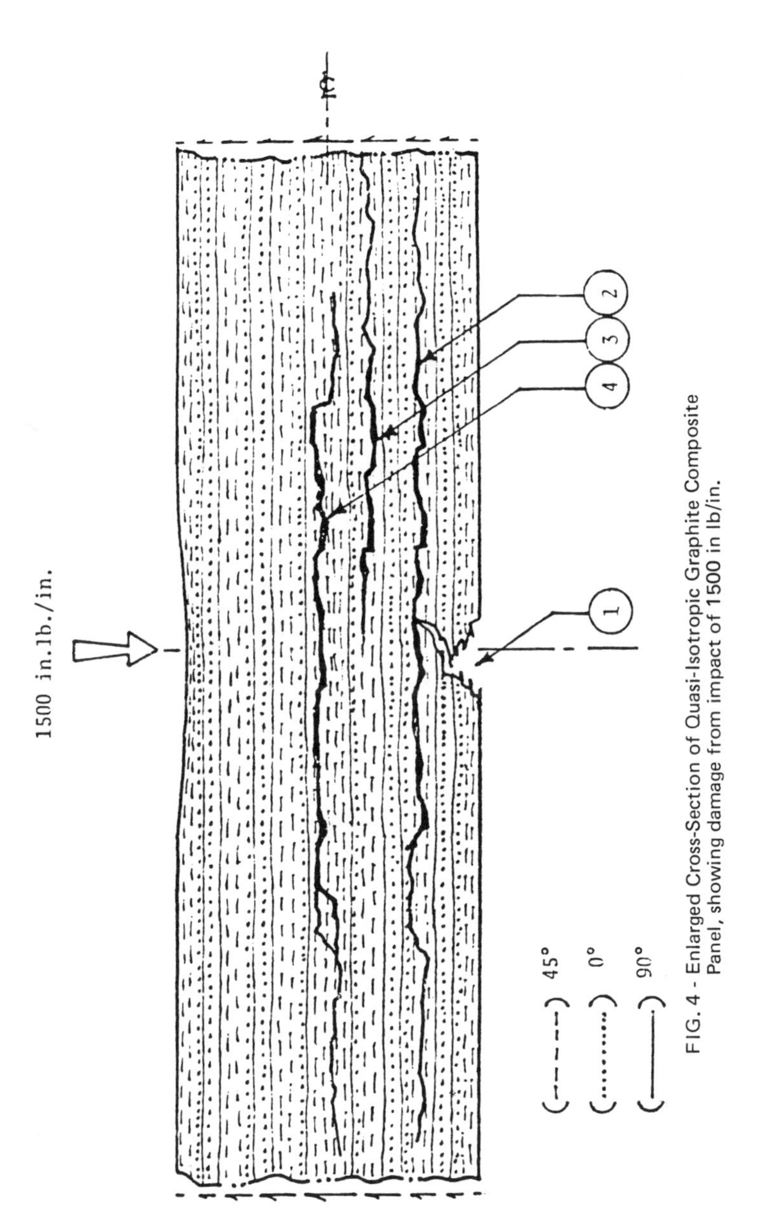

FIG. 4 - Enlarged Cross-Section of Quasi-Isotropic Graphite Composite Panel, showing damage from impact of 1500 in lb/in.

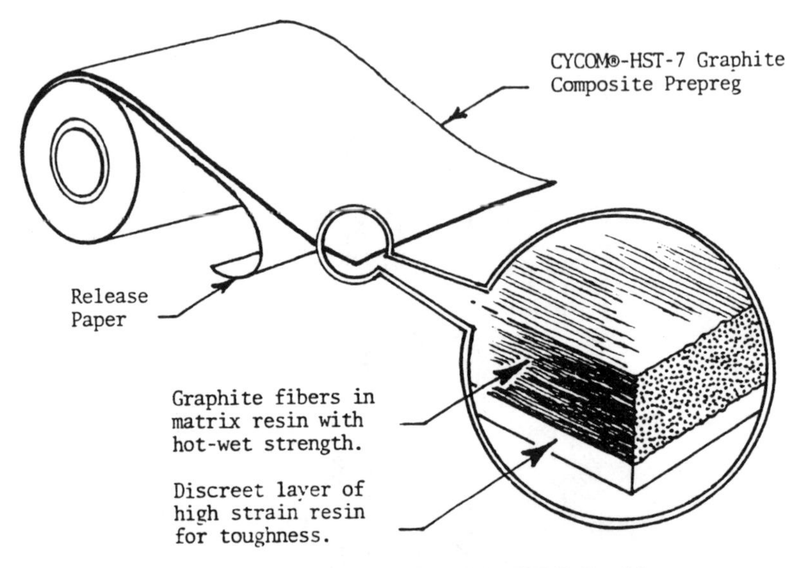

FIG. 5 - Description of CYCOM®-HST-7 Graphite Composite Prepreg.

Conventional Quasi-Isotropic Laminate

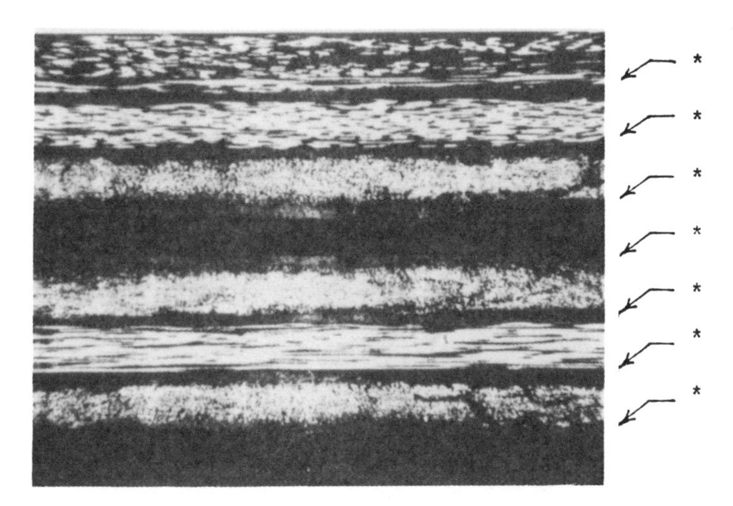

CYCOM®-HST-7 Quasi-Isotropic Laminate

FIG. 6 - Cross-Section Photo-Enlargements of Conventional Mono-Matrix Graphite Composite Laminate and CYCOM®-HST-7 Graphite Composite Laminate. (*) is discreet inter-leaf layer.

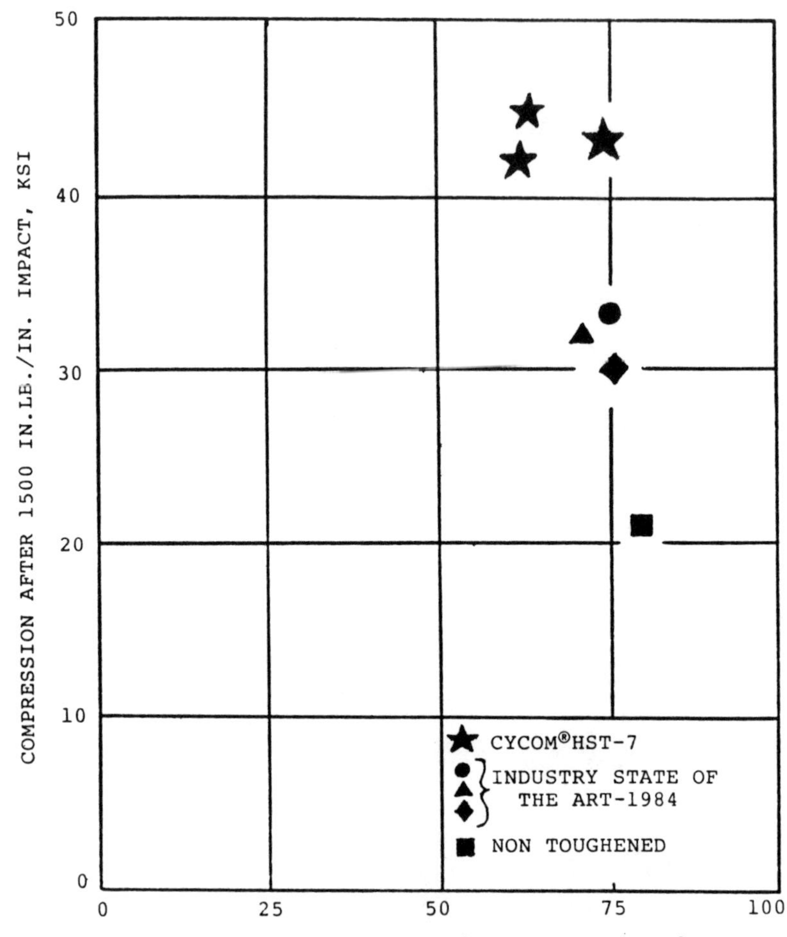

FIG. 7 COMPRESSION AFTER IMPACT VS. HOT/WET COMPRESSION FOR VARIOUS CARBON COMPOSITES

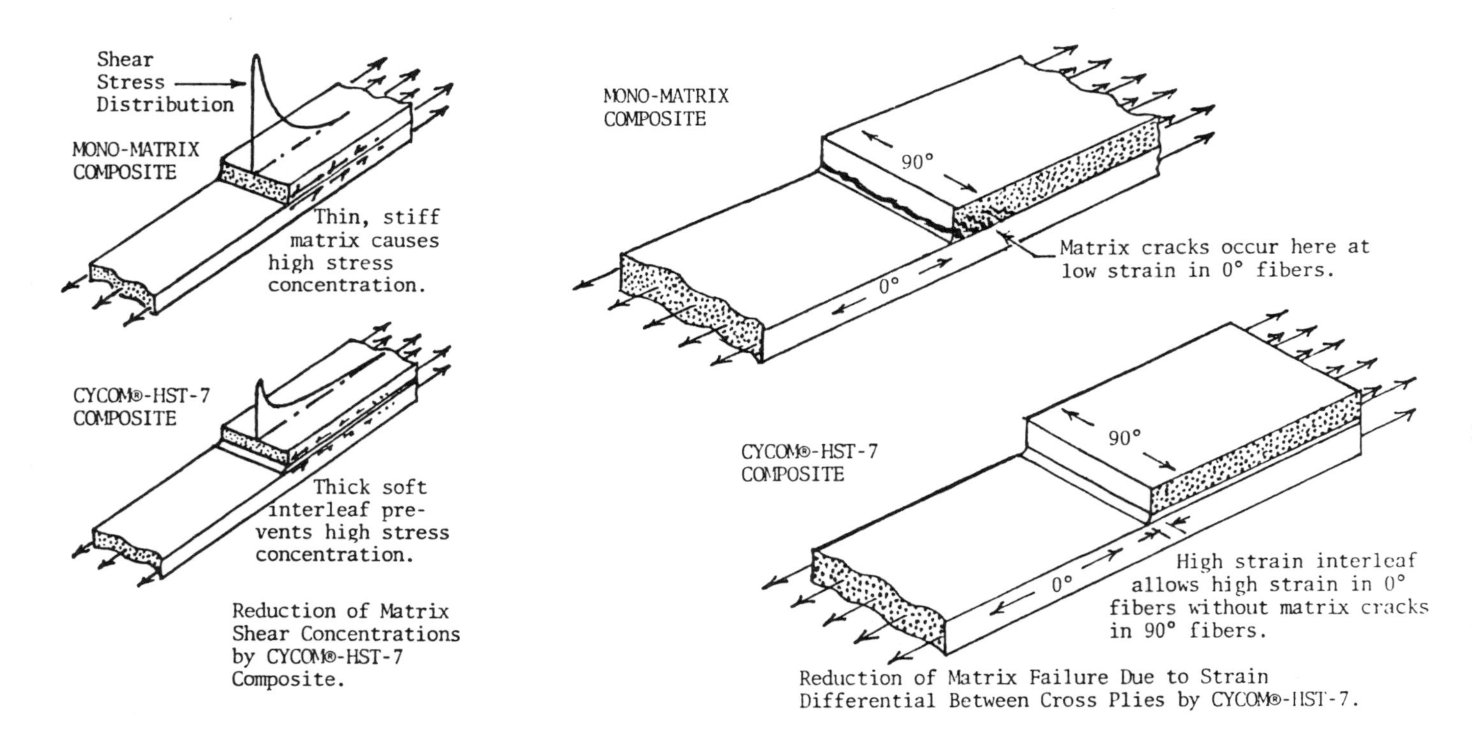

FIG. 8 INTERLEAF STRAIN ACCOMODATION

CARBON FIBER—USES AND PROSPECTS

J.J. Shorter

Chas. A. Blatchford & Sons Ltd.

CFRP is now being used to produce the main load bearing structure of a new artificial limb for the lower extremity, known as the '**ENDOLITE**' system. The use of this material has enabled the designers to dramatically reduce the weight of the limb to a level suitable for the enfeebled geriatric patient whilst being strong enough and sufficiently durable for the most active, heavy weight amputee. The system, including the composite components is now in full production.

BACKGROUND

Artificial limbs for the lower extremity are known to have been in use since 300 BC. Much clever innovation has taken place since this time, but generally speaking wood and leather were the materials used. At the turn of the century we saw the introduction of the light metal limb, and these aluminium alloy limbs were standard issue when the National Health Service came into being in 1947.

1960 saw a departure from light metal limbs of crustacean construction to modular prostheses, where the loads are carried by an inner skeletal frame with an outer soft covering to provide good cosmetic effect. These limbs are produced from stock "off-the-shelf" modules, giving a choice of function at the knee and ankle in a range of sizes to suit the patients' requirements.

These first generation modular limbs, although providing significant advantages in method of production and cosmetic effect, were generally rather too heavy, especially for the more enfeebled amputee.

At an international meeting at Ascot, England in 1966 the clinicians present criticised the weight of these modular limbs, and specified the requirement as being 2Kg for the above knee prosthesis and 1Kg for the below knee prosthesis.

The Company were very mindful of these recommendations when some eight years ago, we drafted the design specification for a new high-tech prosthetic system now known as 'ENDOLITE'. In writing this specification, we took the approach that it should be possible to achieve within the weight limitations, a structure strong enough for the most heavy, active amputee.

We felt that in order to achieve this however, it would be necessary to depart from materials that were generally used for limb construction, and it was at this point that we turned to composites for a possible solution.

CFRP had been considered for a similar application in the U.S.A. where a contract was placed by the Veterans Administration with the Northrop Aircraft Corporation of California in about 1973, but after some progress the project was not considered to be economically attractive by the limb manufacturers in America and was dropped.

Design and development

The two components identified as being most likely contenders for consideration were the main shin structure and the rigid keel of the foot. In both cases the materials used required properties of high specific strength and specific modulus. (See Figure 1).

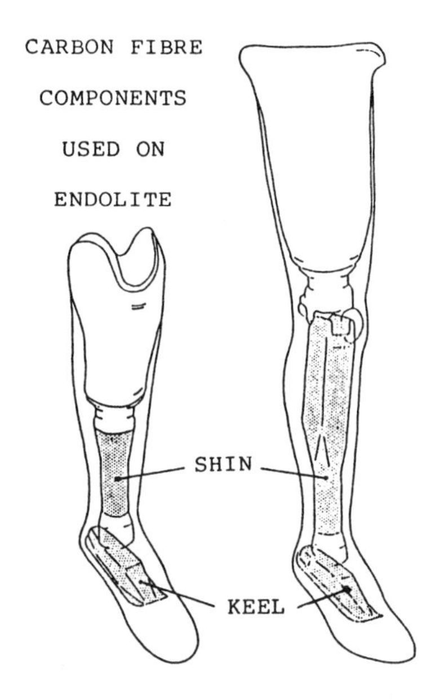

Figure 1. Carbon fibre components used on Endolite

Support for the philosophy of lightweight prostheses, and in particular our choice of components to be manufactured in carbon fibre, came in the form of a recent study from the United States. This work involved experiments on an artificial limb with a shin section having variable inertial properties. Indications were that the weight distribution in the prosthetic shin/foot has a significant impact on gait. The study suggests that future prostheses should be designed to minimise distal shin/foot weight.

We considered various materials during our early investigation on the shin structure including glass and aramid fibres, but as the component is generally subject to compressive loads, carbon fibre reinforcement was our first choice.

The relatively brittle nature of the material, and susceptibility to impact damage, caused us some concern in the early stages, but what better application could there be than one where the structure is eventually 'packaged' within soft cosmetic foam in service.

Initial development work explored the possibility of providing an open patterned winding using tows of carbon fibre wetted out and wound about a former tailored to each individual patient's requirements. The results were interesting but it became clear that without very sophisticated equipment this method was time consuming and the results could be inconsistent.

It was at this stage that we consulted the Atomic Energy Research Establishment at Harwell and after initial discussions placed a development contract with them for the shin. The Company were able to specify accurately many of the design parameters such as envelope size and the shape and form at the proximal and distal extremities to accept the mechanism at knee and ankle. We were also able to quantify the loads to which the structure would be subjected in service. AERE on the other hand, had the necessary materials expertise and fabrication experience.

The project was completed satisfactorily in 1980 and a particular grade of epoxy/carbon fibre prepreg was selected to be used as the basic material. Subsequently, due to close liaison with the prepreg manufacturers, a further grade of cloth/resin combination was adopted in which the resin properties were tailored to meet toughness and processing requirments. A variation of the bladder moulding technique now patented for the application, was used to fabricate the initial shin samples for test and evaluation.

Encouraged by the results obtained for the shin structure, a further programme of development was undertaken in-house to produce the keel of the artificial foot. This had the added problem of being required in a whole range of sizes, thus increasing the tooling requirement.

Initially, the keel was produced in two parts, each being a separate press moulding again in epoxy/carbon fibre prepreg. These top and bottom components were bonded together at final assembly to give a very strong, rigid, hollow box section which could be encapsulated within the flexible material of the foot. Subsequently a UK company, Advanced Composite Components Limited, devised a method whereby the component was moulded in a single process and were thus able to show considerable cost savings, making the project a viable, economic proposition.

With both of these items considerable time and effort was afforded to the interfaces between the CFRP and metallic elements of the components, and as a result, we have accrued a specialised knowledge of compatible high strength adhesives.

Test programme

Since approximately 1970, the Department of Health and Social Security (Scientific & Technical Branch) have subjected all new designs to a formal structural test programme in the laboratory before accepting the component for supply to patients in the UK.

These test methods were as recommended by an international Working Group in Philadelphia in 1977 and relate to both the method of testing and the actual load levels (see Figure 2).

The programme includes static tests to represent peak loads encountered in service and cyclic testing to a lower level to represent commonly repeated loads on the limb during normal use with duration of test set to approximate five years service life.

In the case of static testing, load is applied at a controlled rate, held at the specified level for 30 seconds and removed. If no permanent deformation is evident, load is re-applied and the component taken to failure. According to the mode of failure a factor of safety is applied for acceptance purposes of 2 for brittle failure and 1.5 for ductile failure.

As far as cyclic testing is concerned, the test is accelerated to a sensible degree with duration set at 2×10^6 applications of load at a frequency of between 1 and 10 Hz, dependent upon the sample.

Special tests are also devised to supplement the main programme where appropriate. For example a test with the limb in full flexion to represent a patient kneeling or crouching is extremely searching, as very high moments occur at the knee. This test resulted in a refined shin design modified to remove areas of high stress concentration.

The test programme to date for the Endolite System has consisted of over 160 separate tests over a period of 3½ years, and these in addition to an extensive clinical evaluation under service conditions.

Production

Having successfully completed the test programme for the Endolite System as a whole, the next phase was to put these carbon fibre components into production. Within a small and very specialised market our own requirements are relatively modest. However, the potential world wide for a high tech system with a light weight carbon fibre structure complementing sophisticated functional elements is substantial. The Company therefore considered the need for equipment capable of producing between 5,000 and 10,000 mouldings of each type per annum.

To facilitate production of the shin we have adapted a three platen press, produced for us by Fothergill Rotorway, in order to give an acceptable cycle time with good process control. Similar arrangements have been made for the other components and to date we have approaching 10,000 carbon fibre components in service use in the UK and Europe.

Machining of the mouldings at the interfaces led the Company into investigations relating to the special requirements for cutting methods and tooling for these abrasive materials.

Having experimented with diamond coated, high speed and solid carbide cutters, we have found the use of solid carbide routers at high speeds to be the most suitable. However, amortisation of the cost of cutting tools needs to be constantly monitored as problems with tool life can be costly.

Quality assurance

Quality assurance is one area where we have a considerable dilemma. We have looked at many systems, accoustic emmision, ultrasonics, etc., but to date we are unable to find a single method that is ideal for all applications. We need to be able to detect defects in mouldings and check the integrity of bonded areas without adding significantly to the labour content of an already relatively expensive component. As far as automatic scanning is concerned, this is difficult to tool with such a variety of shapes to accommodate.

Having said this, we have not experienced a single significant quality problem on any of the composite components in service. This is due I feel, to the safety margin applied to the design and to having adopted a philosophy of tight control over raw material used and strict schedules to meet the processing requirements.

We currently carry out weight checks on the finished component in addition to a thorough 100% visual examination where machined surfaces in particular, give a good indication of quality. These measures enable us to continue to produce components with some confidence, whilst seeking an optimised solution regarding the definitive quality assurance procedure to be adopted.

The future

As far as the future is concerned, our own Endolite System will expand to embrace all levels of amputation from the through hip downward, and this will enable further components to be designed in CFRP.

There is also great scope for carbon fibre reinforcement within a thermoplastic matrix where post forming to fit the patient is required, and I have recently seen orthotic components such as calipers in this material, produced in France.

Carbon fibre is also being used routinely in some countries to stiffen laminates produced using hand lay-up techniques and using acrylic resins, although I personally do not see hand laminates of this nature being particularly attractive.

In conclusion may I say that we as a Company have broken new ground with the application of composites technology in lower limb prosthetics. Providing that these materials are used responsibly by maximising benefits and minimising the cost by good design, I am convinced that we will see a substantial increase in the use of CFRP in this aspect of medical engineering.

J Shorter trained as an Aeronautical Engineer with The British Aircraft Corporation, Weybridge. Joined the limb making company of Chas.A. Blatchford & Sons Ltd. as an Engineering Designer in 1967. After holding a number of posts within Blatchfords, was appointed to the Board as Technical Director in 1984.

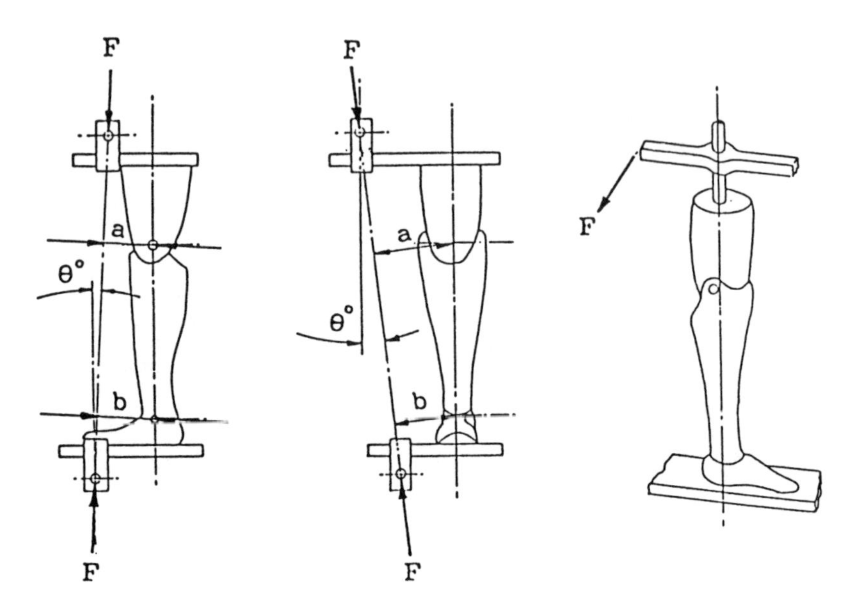

1. COMBINED AXIAL COMPRESSION AND A-P BENDING
2. COMBINED AXIAL COMPRESSION AND M-L BENDING
3. AXIAL TORQUE

	AXIAL LOAD N	AXIAL TORQUE Nm	KNEE BENDING MOMENTS A/P Nm	KNEE BENDING MOMENTS M/L Nm	ANKLE BENDING MOMENTS A/P Nm	ANKLE BENDING MOMENTS M/L Nm
STATIC TEST	2500	35	230	150	250	70
CYCLIC TEST	1350	20	120	80	140	±50

Figure 2. Test method and loading values according to 'Philadelphia' recommendations

Part 4

Performance versus Properties

COMPETITIVE MATERIALS—PROBLEMS OR SOLUTIONS

J. Fray

BAe PLC Woodford Cheshire

Aircraft structural design has been greatly influenced by the periodic introduction of significantly different new materials, although for almost 50 years it has been dominated by the mechanical and physical characteristics of aluminium alloys. However, several new materials which offer very attractive packages of properties are now challenging the aluminium alloys for a share of the aircraft structural market and the advent of these materials seems likely to influence not only aircraft design but also the structure of the industry itself.

MATERIALS IN QUESTION

This paper considers competitive materials in the context of the structure of present-day and future aircraft. However, the picture presented is not untypical of other sectors of the transport industry with some differences in emphasis, and is essentially one of an increasing number of new materials that are being offered to designers as superior (but in almost all cases more expensive) alternatives to the traditional materials.

Designers are of course always on the look-out for materials which offer better solutions to their many design problems, and the progressive development of the aeroplane over the last seventy years or so clearly bears witness to the influence of new materials and new materials-related manufacturing processes on structural design.

It is in fact quite difficult to define the term "better" in the context of alternative materials. In aircraft design there is considerable emphasis on the reduction of weight, since this is directly reflected in the operating performance (and hence in the operating cost) of the aeroplane; but other factors such as cost of manufacture, reduced maintenance requirements and a longer service life are also very important considerations.

A reduction in structure weight can result from:

a) Higher material strength - weight and/or stiffness - weight ratios.

b) Higher values of $E^{1/2}/\rho$ - the material parameter which defines the efficiency of a structural member designed by stability.

c) A better fatigue performance - ie more load cycles to the onset of cracking and/or a slower rate of crack propagation.

d) Greater toughness, ie higher static strength in the presence of a crack, leading to longer periods between mandatory inspections.

e) A more statistically valid data base.

f) The elimination of joints.

Lower costs, in terms of initial manufacture and subsequent maintenance, can come from:

a) Lower material cost.

b) Better material utilisation.

c) Easier fabrication.

d) Fewer components.

e) Greater resistance to corrosion.

In addition to these straightforward technical considerations there are other rather less quantifiable factors such as:

a) Crashworthiness in terms of energy absorption, fuel containment, smoke emission and toxicity. Also resistance to bird and lightning strikes, electromagnetic compatability etc.

b) Customer appeal in terms of good appearance (passengers may prefer to see a smooth external suface rather than the rippled and fragile-looking surface of a thin metal structure) and the real or imagined benefits of "high technology".

c) Better aerodynamic performance as a result of precise contour control and very smooth surfaces.

d) The expertise, traditions and the like of the particular company, the facilities available to the company and the cost of change.

e) The instincts of the designer for what will provide the best answer to the sum of the individual problems as he perceives them.

All materials offer a unique balance of properties and the designer has somehow to decide what is best for his particular purpose in the situation in which he finds himself. This decision has to be based upon a combination of calculation and intuition.

MATERIALS IN COMPETITION

Until quite recently the choice of material for aircraft structures was relatively easy. Unless the operating temperature of the structure was over about 120°C (which was seldom the case) aluminium alloy was the natural choice, with copper-containing alloys being preferred for fatigue-critical applications and zinc-containing alloys for applications where the higher strength of these alloys could be used to advantage. If the temperature was too high for aluminium alloy the choice fell to titanium, either commercially pure for heat shields and the like or probably a 6Al-4V alloy for more highly loaded applications. Nickel alloys, and some steels, have proved to be very useful for parts required to work at particularly high temperatures, and very dense structures (ie components of limited dimensions carrying high loads) have also been made from steel.

Magnesium has had a somewhat chequered career in aviation. Although it is much lighter than aluminium, the readiness with which magnesium corrodes (sometimes in the stores as well as on aeroplanes) if it is not protected poses risks which designers are seldom prepared to take. If this problem could be overcome metallurgically there could be a very big aviation market for magnesium alloys.

Many of the traditional structural materials are now being challenged by a whole range of new materials, some of which are available for immediate use although the majority are to varying degrees in the future. These new materials (and new material-based manufacturing processes which extend the scope of some traditional materials) include:

a) New materials now available:

Second generation carbon and aramid thermosetting composites.
Superplastically formed/diffusion bonded titanium alloys.

b) Materials that will be available in the near future:

Aluminium - lithium alloys.
Powder alloys.
Third generation carbon and aramid thermosetting composites.
Thermoplastic structural composites.
Superplastically formed aluminium alloys.

c) Materials likely to be available in the longer term:

Aramid reinforced aluminium laminate (ARALL).
Metal matrix composites.
Advanced aluminium (and possibly also magnesium) alloys produced by novel metallurgical techniques.
Roll-bonded aluminium alloys.
Superplastically formed/diffusion bonded aluminium alloys.

Aluminium alloy has clearly offered an excellent balance of properties for aircraft structures. The ready acceptance of aluminium alloy by the aircraft industry during the 1930's can be clearly seen by looking at the development of aircraft prototypes over this period. What is it that aluminium alloy does in fact offer? This can perhaps be summarised as follows:

a) Its density is about right for most components in that it is not so low that an efficient structure cannot be conveniently fitted into the space available, and not so high that producing a stable compression structure is not too difficult or expensive. In almost all cases even the thinnest panel can be adequately stiffened by a simple stringer.

b) It can be stretched, bent and otherwise manipulated on inexpensive tooling with the aid of low-temperature heat treatments. It is very easily machined and scrap material finds a ready market.

c) It has a fairly good resistance to corrosion (depending of course on the basis of comparison) and it can be given a high standard of protection by straightforward treatments.

d) It is relatively insensitive to notches so that there is a very good chance that a crack will be discovered well before it becomes dangerous.

e) It can absorb impact energy by plastic yielding. In fact conventional skin-stringer-frame structure absorbs energy very well. Minor damage can be easily seen with the unaided eye and is seldom critical.

Titanium on the other hand is a relatively difficult material to form and to machine, although it is readily welded and is free from in-service corrosion. Furthermore the material thickness required is often small and hence extensive stiffening is usually necessary. However, some titanium alloys do exhibit excellent simultaneous superplastic forming and diffusion bonding properties which can be used to good advantage in the manufacture of complex components at a significant cost advantage. The economic advantages of the SPF/DB process (particularly when a large number of identical components are required, or a complex built-up assembly can be replaced by a single item) can be such that the designer may choose to use titanium for applications for which it would otherwise be totally unsuited.

Although perhaps not too inordinately expensive when compared with some of the newer materials, titanium alloys nevertheless cost some 7 to 15 times more than aluminium alloys depending upon form - with thin sheet being the most expensive. This disadvantage has to be overcome by savings in fabrication costs and/or by imparting some operating advantage to the aeroplane.

Perhaps the most exciting development at the present time is however aluminium-lithium alloys since these alloys offer an 10% decrease in density together with a 10% increase in stiffness. Hence there is the prospect of an immediate 10% reduction in structure weight being obtained by direct substitution (ie without any dimensional changes) for the existing material - assuming of course that there are no problems of load redistribution due to the increased modulus. By taking advantage of the higher modulus, weight savings of nearly 20% are theoretically possible in stiffness designed structure - and about half the structure of a modern aeroplane is designed to stiffness criteria.

All this can be obtained from a material that can be fabricated by conventional methods and on existing equipment. That is of course the promise - the practice may be somewhat different since nothing is ever quite so straightforward. Indeed, although it is still very early days yet, there are fears that the crack propagation resistance of Al-Li alloys may limit their range of application, that there may be problems in maintaining the lithium content when rolling into thin sheet (although this may be overcome by treatment in close coils) and there may be practical difficulties such as segregating the highly reactive lithium - containing scrap. Furthermore, the slow natural ageing characteristics may be found to be a disadvantage in fabrication, as may the different reponse to standard anticorrosive treatments such as anodising. An increased tendency to corrode may require more attention to temporary protection on the shop floor and to more permanent protection in service. These differences in behaviour will undoubtedly be reflected in somewhat higher manufacturing costs over and above what seems likely to be about a 3-times premium on raw material cost.

Nevertheless, the longer term prospects for aluminium-lithium alloys appear to be very good.

It seems to be an inescapable feature of the metallurgy of aluminium (and broadly that of other metals also) that the price of increased strength is reduced toughness and/or reduced crack propagation resistance. A novel solution to this problem is being offered by the ARALL concept. This synergistic combination of aramid composite and aluminium sheet (although very complex from the point of view of structural analysis since it offers almost limitless possible combinations of parameters) may provide a means of utilising aluminium alloys which would be very attractive but for their poor resistance to crack growth. The ARALL concept is by no means limited to aluminium alloys (although a successful aluminium-based material is the first target) and may indeed provide a means of overcoming crack propagation problems in other high performance metals.

Mention of ARALL brings the subject back to composites. Metal matrix composites, typically with an aluminium alloy matrix and fibrous or particulate silicon nitride as the "reinforcing" phase, are now available for the manufacture of test items and are showing good promise for both strength and stiffness designed structure.

Although undoubtedly much more difficult to turn into useful hardware than aluminium-lithium alloys, metal matrix composites offer considerably greater weight saving potential with specific strength (E/ρ) and specific stiffness (σ/ρ) advantages over conventional alloys of three times and four times respectively in favourable cases. Hence, although it is early days yet (particularly in Europe, although the situation in the USA is more advanced) metal matrix composites are clearly worthy of some priority in the increasing competition for R&D funding.

The interest in carbon fibre composites still remains strong and some large items are now being fitted to production aircraft with many more in various stages of development. Indeed it would be fair to say that carbon fibre composites have now become an accepted range of structural materials that stands on its own merits - even though there is still considerable scope for improvement in the materials on the market today.

What are then the advantages of carbon fibre composites in the context of today's (and tomorrow's) alternatives? Firstly the specific strengths and specific stiffnesses of CFC laminates compare very favourably with the equivalent properties of the alternative materials, and particularly so when the load paths for which the laminates are designed are clearly defined, but it is perhaps the other factors which offer most advantage in practical applications. These factors include:

a) Freedom from corrosion. This is a bigger advantage in many cases than may sometimes be realised, particularly for aircraft (and helicopters) that may be operated in marine enviroments. Hence closed box structures (which can be very efficient structurally) become acceptable.

b) Freedom from fatigue under tension loading - at least at current allowable strain levels. However, impact damage (barely visible or otherwise) can propagate rapidly and catastrophically under compression loading.

c) A greatly reduced need for mechanical joints - with consequent weight and cost savings.

d) The ease with which accurate and smooth shapes can be repeatedly produced on relatively low cost tooling.

There are clearly also disadvantages, but these are tending to lose their significance as experience grows. Nevertheless, thermosetting composites in particular still suffer from the fundamental problem of poor resistance to the propagation of delamination damage (which is particularly relevant to structures loaded in compression and shear) and this perhaps remains the most urgent development area.

Some of the earlier areas of concern have proved to be of little consequence in reality. For instance, experience has shown that the fire containment characteristics of carbon fibre composites can be much better than those of aluminium alloys. Although a flame impinging on a composite panel will cause considerable local damage, the charred composite will contain the flame long after the point at which an equivalent aluminium alloy panel would have melted. There is indeed evidence to show that a carbon fibre composite fuselage could be designed to offer a very much greater degree of protection to passengers in the event of a major fuel fire than an equivalent, and possibly much heavier, aluminium alloy fuselage.

Other features about which designers are now more relaxed, or for which effective solutions have been found include:

a) The use of mechanical fasteners

b) Variability

c) Lightning strike

d) Corrosion in aluminium - carbon joints

e) Quality control

f) Repair

MATERIALS IN COLLABORATION

How will this bewildering variety of new materials come together in structures that have to earn their keep in a competitive world? This will obviously depend upon individual problems, preferences and prejudices as well as the relative progress of the competing technologies. Nevertheless I should like to offer some personal views on how the various alternative materials may be used in the aircraft structures of the future.

Aluminium Alloys

Now that the aluminium producers have reacted with some enthusiasm to the challenge of composites the early introduction of useful new aluminium alloys onto the commercial market can be anticipated with confidence.

Many of the new alloys will be produced by novel techniques in order to overcome the limitations of conventional billet casting, but it is nevertheless likely that at least some of these alloys may suffer from poor resistance to crack propagation. Hence it may be difficult to utilise their full strength potential in structures (eg lower wing covers) that are subjected to tensile fatigue loading. Improved aluminium alloys will therefore probably find most favour for compression loaded structure (eg upper wing covers), for large monolithic items (eg forgings and castings) and for much of the general sub-structure (frames, ribs etc) of aircraft other than the small combat types where the special requirements referred to below apply.

The possibility of developing reliable and cost effective superplastic forming (and also simultaneous diffusion bonding) techniques for high strength aluminium alloys further increases the probability that aluminium will remain the principal aircraft structural material for the forseeable future.

ARALL

The ARALL concept may, in the longer term, provide an answer to the crack propagation resistance problem exhibited by current as well as possibly by future high strength aluminium (and other) alloys. However there is a great deal of work to be done to develop ARALL from its present laboratory status into a fully characterised, commercially available material. Nevertheless ARALL could eventually find extensive use for areas of the airframe designed by fatigue considerations, the skins of pressurised fuselage and lower wing covers being prime examples. The unique behaviour of ARALL may however pose some very interesting questions regarding the safety of structures which may contain a multiplicity of cracks.

Titanium

The "hot" areas of both military and civil aircraft are likely to remain the main applicatons for titanium although the continued development of superplastic forming - diffusion bonding will encourage the use of titanium as an alternative to aluminium for items that can be formed in one piece. Doors are an example of such a possibility.

Metal Matrix Composites

The future of metal matrix composites in aircraft structures is difficult to forecast at the present time although the potential for both the isotropic particulate reinforced material and the anisotropic continuously reinforced material (with aluminium alloy as the matrix in both cases) appears to be considerable in the longer term. Again, resistance to crack propagation might restrict the application of particulate-reinforced, if not also the continuously-reinforced materials, and hence one is tempted to think of an ARALL type laminate utilising a reinforced aluminium alloy as the metal element. Perhaps such materials will become commonplace in the early 2000's.

The strength and stiffness advantages of metal matrix composites can be considerable, particularly if one can think in terms of multiple laminates with continuous reinforcement orientated in the manner of current non-metallic composites.

Carbon Fibre Composites

It would seem fairly certain that the use of carbon fibre composites will continue to expand as confidence in their behaviour grows. The wings and empennage of combat aircraft and the control surfaces (including perhaps the fins and tailplanes) of transport aircraft are likely to be major applications.

The wings of combat aircraft are an interesting illustration of the use of carbon fibre composites to solve a particular set of design problems - which are essentially to fit a minimum weight structure into what is in effect a shallow, closed box and at the same time to control the bending and twisting of this box in a prescribed way under the action of the aerodynamic and inertial forces. This has become known as "aeroelastic tailoring". The freedom with which one can build up a carbon fibre structure from the basic unidirectional prepreg into large joint-free items has proved to be a major advantage in the design of wings for combat aircraft.

CONCLUSIONS

It would seem fairly certain that the future will see a major swing away from what have been virtually all-aluminium aeroplanes to aeroplanes built up from several substantially different types of material.

The change of emphasis in materials may not only lead to new forms of aircraft structure but may also have a significant influence on the structure of the aircraft manufacturing business itself, with greater reliance being placed on specialist suppliers and closer supplier/purchaser relationships. Whether the diverse materials will be happy bedfellows however remains to be seen.

ACKNOWLEDGEMENTS

The author would like to thank his many colleagues in British Aerospace for their invaluable assistance in the preparation of this paper and the Directors of British Aerospace for permission to publish it. The views expressed are nevertheless his own and do not necessarily represent Company policy.

CARBON/CARBON DEVELOPMENTS FOR AIRCRAFT BRAKES

T. Wells

Dunlop Aerospace

Paper not available at press time.

CARBON FIBER COMPOSITE RECIPROCATING GUIDE BAR

Donald R. Lovell

Materials Consultant
HiPerTech, 'Reapers Croft'
Chilcompton, BATH BA3 4HB

A CF/epoxy square section tube made in three sections each about 2 m long replaced a similar steel tube as a traverse bar on a textile spinning machine. It reciprocates longitudinally over 75 - 200 mm driven by a cam and follower at one end at 750 cycles/minute. The steel bars failed by fatigue in less than 4×10^8 cycles (one year). The composite bars have achieved 6×10^8 cycles without failure and are still operating.

The design, fabrication and testing of the composite bars are described and the cost effectiveness discussed.

INTRODUCTION.

A spinning machine for synthetic textile yarns has on each side a 12.6 mm square section bar made in several sections each 1.5-2 m long to a total length of 5.5-9 m, according to the particular version of the machine. These 'traverse bars' carry thread guides at 150 -300 mm intervals and pass through bearings at 320 - 440 mm spacing. The bars reciprocate longitudinally at 750 cycles/minute over a distance variable between 75 and 200 mm and are wetted by a spray of spin-finish liquor (of confidential and unknown composition) from the yarns.

The hollow steel traverse bars break, probably due to fatigue failure caused by the vibration wave transmitted down them from the driving cam at one end. The machines operate continuously for 50 weeks each year and the life of the metal bars has been less than one year. In one year, 8400 hours, the bar undergoes 3.78×10^8 cycles.

By the use of carbon fibre composite (CFC) it was hoped to increase the speed by 10% or more and increase the life to at least 5×10^8 cycles (10,000 hours). If successful, there was at the time the work started, predicted to be a potential requirement for 160 bars immediately and a further 800 longer term.

DESIGN OF COMPOSITE BAR.

The basic design philosophy was to provide a prototype which would not fail during trials even if it did not have the optimum performance. After this had engendered confidence in CFC material, an optimisation of the design could follow.

Matrix Resin.

The resin had to withstand the effect of the spin-finish liquor which was found to attack some epoxy resins causing swelling (Shell 'Epikote 828/NMA/BDMA system). Laboratory tests suggested that the polyester Atlac 382-05A and a vinyl ester Derakane 470-5 were not attacked. In the event, the fabricator used a proprietary epoxy system which performed satisfactorily.

Fibre Type.

The properties used in the design of the four types of carbon fibre considered and their composites are compared with other materials in Table I. Boron fibre was a possible alternative to carbon fibre but was rejected because it was not readily available and experience of fabricating composites with it was lacking in the UK.

Calculations of the stress in the bar at various speeds and throws (distance travelled between reversals) were made for three types of carbon fibre - UHM, ultra-high modulus; HM, high modulus; VHS, Very High Strength - and for steel, using fibre modulus values of 483, 345, 206 GPa respectively. The results are shown in Table II from which it can be seen that the minimum stress was obtained using HM fibre, although the difference between UHM and HM was small. As HM was more readily available and cheaper it was selected. For all CFC materials as unidirectional, 60% fibre volume fraction composites, the stress was 25 - 30% of that in steel.

Joints between sections.

It was not possible to fit a single composite bar 6 - 9 m long to the machine because of site limitations, although this would clearly be desirable. The metal sections were therefore reproduced. These ended in conical joints which were clamped together. A conical end was formed on the CFC tube by belling out the fibres and adding circumferential plies of fabric. No attempt was made to replace the metal clamps for the initial trials, although it was intended to substitute lighter components later.

Bearings.

Consideration was given to using carbon, carbon-filled PTFE, carbon fibre/epoxy or PTFE-filled metal bearings in place of the oil-impregnated sintered metal now used. As it was important that the trials of the composite bars should not be interrupted by bearing failure or wear of the composite in the bearings, it was decided to fix thin metal coverings over the CFC bars for the length which would run in each bearing and use the existing bearings. This was in line with the design philosophy and was satisfactory.

Thread Guides.

The thread guides consist of acetal clamps carrying an alumina guide which are fixed to the bars by a single screw so that their spacing can be adjusted quickly. These sometimes slide along the metal bars and an improved method of fixing was required. This was achieved by serrating the surface of the CFC bars.

Dimensional Tolerances.

The metal bars were 12.60 $\pm$ 0.05 mm square outside x 1.27 mm wall thickness.

The bearings were 12.66 mm square x 33.4 mm long.

The straightness tolerance for the metal bars was 0.167 mm/m. It was appreciated that achieving this tolerance on a composite bar would be expensive and as the maximum allowable variation between a pair of bearings was 1 mm, in order to control the distance between the thread guide and the 'cheese' of wound fibre, an initial straightness tolerance of 1 mm/m was set. No limit was placed on twist in the composite bar and this proved a problem on the first mouldings as described later.

Fibre Arrangement.

Although the calulations to select the fibre type were made for unidirectional reinforcement, the design philosophy indicated that some transverse fibres should be incorporated. A combination of HM unidirectional and VHS woven fabric was used which gave about 84% of the fibre along the length of the bar.

FABRICATION.

Four processes were considered.

PROCESS	ADVANTAGES	DISADVANTAGES
Pultrusion	Cheapness Low tooling cost	Poor tolerances on outside dimensions and straightness U/D reinforcement only No integral joints
Hot Press Moulding	Good control of dimensions and tolerances Transverse fibre possible Integral joints and range of lengths possible	Requires 2 m long press Requires prepreg
Internal Bag	Good outside dimensions Transverse fibre possible Integral joints and range of lengths possible	Inside shape circular Consolidation in corners may be difficult Requires prepreg
Resin Injection	Good control of dimensions Transverse fibre possible Integral joints and range of lengths possible	Limited range of resin viscosity acceptable

The internal bag method was selected since pultrusion could not produce the integral joints required; resin injection along a 1.5 mm wall thickness up to 2 m long was likely to be difficult and hot press moulding was a more expensive process.

A female tool 12.6 ± 0.05 mm square x 1.924 m long was made of steel with inserts so that it was also capable of making tubes 1.759 and 1.626 m long,all with a tolerance of ± 1.6 mm. This cost only £ 800. One set of these three lengths formed one traverse bar 5.31 m long. Each section had a different combination of conical ends. The drive end cone had a projection on the face; the tail end section had only one conical end; the centre length had cones at both ends matching those with which it mated on the drive and tail sections.

Preimpregnated unidirectional woven tape of High Modulus carbon fibre (Grafil HMS, Carrfibre CR111) and lengths of woven Very High Strength carbon fibre fabric (Grafil XAS, Carrfibre CR 114) were formed into a tube containing two

two plies of fabric and the remainder unidirectional, around a rubber bag. This assembly was placed in the mould and the rubber bag inflated whilst the resin was cured at 120°C.

The end cones were formed by belling out the unidirectional HMS fibre and adding additional lengths of fabric (Fig 1). The continuity of the HMS fibre ensured that the stresses were transmitted into the conical joints without an adhesive layer which could be a source of weakness. The flat end faces for each cone were press-moulded separately and bonded on.

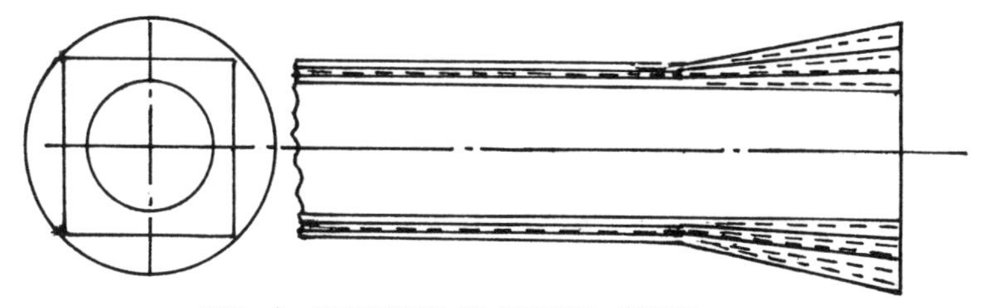

FIG. 1. FORMATION OF CONICAL JOINTS.

Along the parts of the bar where thread guides are fixed, the surface was serrated to prevent the guides sliding as they had done on steel bars.

Where the bar would pass through bearings, metal shims were inserted in the surfaces. These were L - shaped to fit round the corners of the square section and provided a metal surface against which the oil-impregnated sintered metal bearings could run.

Because only a few sets of sections were to be made initially, the first mould was made as cheaply as possible. This may have been the reason for these first sections having some twist in the mouldings which made it necessary to spend extra time during the fitting of the composite bars to the machine for trials to get the bearings to run satisfactorily. In spite of this, when fitted, these prototype traverse bars performed well.

OPERATING TRIALS.

One machine was fitted with composite traverse bars on each side. It has run for over one and a half years (12,600 hours) without failure or problems, at the standard speed of 750 cycles/minute i.e. for approximately 6 x 10^8 cycles, almost twice the life of a typical steel bar. It was still operating satisfactorily when the trial was stopped. No trials have been made at higher speeds.

COST.

The price of a metal section at the time this work was done was £ 50. The prototype composite section was £ 113. Some reduction in this cost was expected when production quantities were fabricated.

If a CFC bar had twice the life of a steel bar, as has now been indicated, there would be a definite saving in overall cost because the loss of production during the replacement of a broken steel bar would be eliminated.

FURTHER DEVELOPMENTS.

From Table II it can be seen that the maximum stress in a composite bar for the increased speed of 950 cycles/minute, a 26.6% increase in speed, is only 5,151 lb/sq.in.(35.5 MPa). As the strength of the material, from Table I allowing for 86% of the fibre content being along the length, is 1303 MPa, the calculated stress at the highest speed is only 2.7% of the strength, which suggests that this 26% increase in output could be achieved without reducing the life of the CFC bar. It would probably be necessary to replace the steel connecting link between the cam follower and the traverse bar with a CFC link to prevent fatigue failure of this component.

EPILOGUE.

The Company started this investigation in 1971 and, after the first supplier failed to make progress, contacted another in 1976 who subsequently closed down. So in 1979 the writer was asked to carry out a feasibility study which resulted in prototype bars going on test in 1980. After some modifications, running trials on the final version commenced early in 1982 and continued until 1984.

Now these spinning machines are being replaced with a new design which does not have traverse bars.

If the company had used the services of a consultant from the beginning they could have had ten years of increased output as well as lower maintenance costs.

CONCLUSIONS.

The design philosophy, which was to produce a prototype which would not fail during trials even if it did not have the optimum performance, was justified.

The stress in the CFC bar was 25 - 30% of that in the steel bar. The strength and vibration damping factor were considerably higher than that of steel.

The low stress and high strength suggest that a 25% increase in speed and output could be achieved, although this has not been confirmed by tests.

At standard speed the composite bars have had at least twice the life of the steel bars when the trial was stopped. Their cost was about twice that of steel.

ACKNOWLEDGEMENTS.

This feasibility study was carried out for ICI Fibres Ltd, Gloucester, and their permission to publish this paper is acknowledged with gratitude.

The prototype bars were fabricated by Advanced Composite Components Ltd,Heanor, Derbyshire, and Mr Roger Sloman is thanked for the information provided on this part of the project.

MATERIAL.	MODULUS (E) GPa	STRENGTH GPa	DENSITY (d) g/cm^3	DAMPING FACTOR $\sqrt{E/d}$
Fibres.				
Carbon UHM	483	2.27	2	-
Carbon HM	345	2.53	1.8	-
Carbon VHS	206	3.16	1.75	-
Carbon HS	172	2.53	1.75	-
Boron	414	4.60	2.3	-
Aramid	124	3.62	1.45	-
Composites. (V_f= 0.6: u/d)				
Carbon UHM	290	1.38	1.68	13.1
Carbon HM	207	1.52	1.56	11.5
Carbon VHS	124	1.90	1.53	9.2
Carbon HS	103	1.52	1.53	8.2
Boron	248	2.76	1.86	11.5
Aramid	76	1.38	1.38	2.8
Metals.				
Steel (ISO typeC35)	206	0.42(yield)	7.8	5.1
Aluminium alloy(6061-T6)	69	0.26	2.56	3.8
Titanium alloy (6Al-4V)	112	0.98	4.45	5.6

TABLE I. COMPARISON OF PROPERTIES OF FIBRES,COMPOSITES & METALS.

CYCLES/MINUTE (Cam speed)	THROW (Rev. angle)	MATERIAL	STRESS lb/sq.in.	MPa
780	17	Steel	12,790	88.2
		UHM/epoxy	3,696	25.5
		HM/epoxy	3,648	25.1
		VHS/epoxy	4,552	31.4
950	17	Steel	20,077	138.5
		UHM/epoxy	5,511	38.0
		HM/epoxy	5,151	35.5
		VHS/epoxy	5,996	41.3
780	27	Steel	7,148	49.3
		UHM/epoxy	2,250	15.5
		HM/epoxy	2,189	15.1
		VHS/epoxy	2,642	18.2
950	27	Steel	10,789	74.4
		UHM/epoxy	3,306	22.8
		HM/epoxy	3,041	21.0
		VHS/epoxy	3,418	23.6

(Stresses were originally calculated in lb/sq.in.)

TABLE II. CALCULATED STRESS IN BARS OF DIFFERENT MATERIALS.

Donald R Lovell graduated in Chemical Engineering from Loughborough College. He worked for the A.P.V. Co. Ltd and the South Eastern Gas Board before joining the Morgan Crucible Group in 1957. After some years in charge of the Furnace Design section in Morganite Research & Development Ltd he became Manager of the Process Engineering Department and then the New Products Department. When the manufacture of High Modulus Carbon Fibre was started he was successively Development, Production, Quality Control and Applications Manager in Morganite Modmor Ltd. Since 1978 he has been an independent Materials Consultant on high performance fibres and composites trading under the name HiPerTech. He compiled the 'World-wide Carbon & High Performance Fibres Directory now in its 3rd Edition, and contributes to and distributes for Composite Market Reports Inc, the Key Personnel Lists for USA and Europe for the Advanced Composite Materials industry.

ARAMID FIBERS FOR ADVANCED COMPOSITES FROM RECREATIONAL TO MILITARY APPLICATIONS

Karlheinz Hillermeier

Enka AG
Wuppertal, Germany

The most important reinforcement properties of aramid fibre filament yarns are being considered in comparison to the new carbon fibre filament yarns. New test methods on the properties of aramid and some differences between carbon fibres and aramid are being discussed. The emphasis is e.g. on modulus, moisture absorption and ballistic properties. These properties of the fibre are used to show the fields of application. The wide range in which the aramid fibre is employed is shown by a number of examples.

INTRODUCTION

On taking stock after a period of more than ten years that aramid has been used as a reinforcing material we can say that this fibre has made its way. On its path to a high-performance reinforcing material for organic matrix systems it has not always followed a straight, but mostly a logical course. From its property profile (1) it became evident at a very early stage that in some respects this fibre is a very different material from textile glass or carbon fibres and that its properties are often even opposed to the latter materials. This was bound to lead to immediate competition with the two other major groups of reinforcing fibres, resulting in fruitful selective research and development. In view of the increasingly differentiating approach to high-performance fibre composites on an even higher level, this competition, especially with carbon fibres, is accompanied by mutual stimulation in terms of technical application development.

Development of properties of aramid fibres and carbon fibres

Developments in these two groups of high-performance fibres varied considerably.

The development of aramid fibres in the early seventies (2, 3 and 4) set new standards for organic reinforcing fibres. It was only in the last few years that aromatic polyether amide (5) was added to the paraphenylene terephthalamides, but for the time being, this fibre does not play a significant part as a reinforcing material. Paraphenylene terephthalamide fibres, including Twaron HM, have met with growing demand in recent years. The product was developed further in terms of process improvements and a systematic, although moderate, increase in filament strength.

Progress was achieved not only by improving the manufacturing conditions, by optimal use of the fibre in specific applications (6) but also by the determination of fibre properties under the aspect of actual application conditions. The latter also led to higher values in the determination of strength, owing to improved test conditions, as shown in Fig. 1.

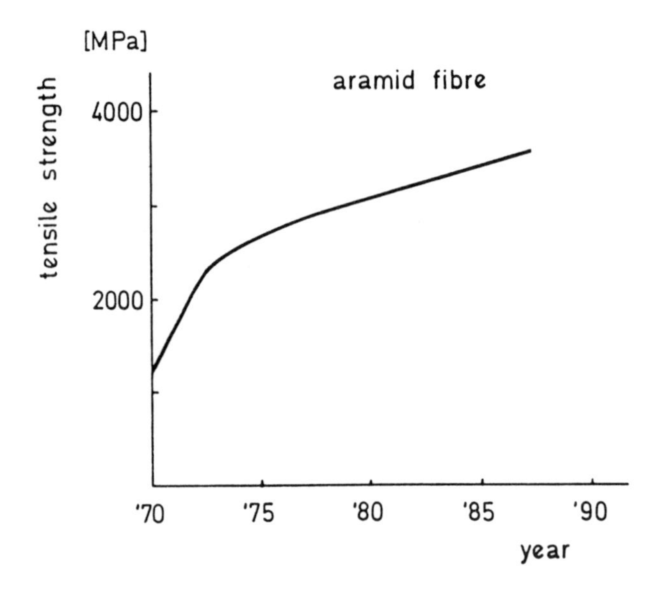

Figure 1 Relationship between tensile strength of aramid fibre and developing period

During the past ten years carbon fibres experienced a steep upturn which has obviously not yet come to a halt.

Figure 2 shows how the properties of the Tenax carbon fibre developed in the two important modulus categories. High-tenacity carbon fibres having a modulus of 235 GPa were developed mainly in the years from 1977 to 1982. The development of intermediate carbon fibres of a modulus of about 300 GPa was initiated in the eighties and is still in progress. The curves of Fig. 2 represent the development of these two fibre types. The first group has been further developed systematically since the mid-seventies. The latest maximum values, however, are not yet reflected by corresponding sales, because tough matrix materials that could be subjected to such strong deformations without damage to the composite are not yet available. This is one of the reasons why the second group of intermediate carbon fibres has made such rapid progress during the last two years from type HM 28 to IM 400 and IM 500 and finally to IM 600 having a modulus of about 300 GPa: in this fibre category the deformations produced by the same tension are less pronounced, which means that the matrix and the interface are subjected to reduced stress. The third and last group comprising the high modulus fibres from HM 40 to HM 50 was not the focal point of development activities during the last decade because of their lacking market potential. But since 1983 the R + D activities have become important because of the growing markets in national and international satellite programmes for telecommunication and military tasks. This will be continued in the coming years when high-modulus fibres with increased strength values are expected to be further developed.

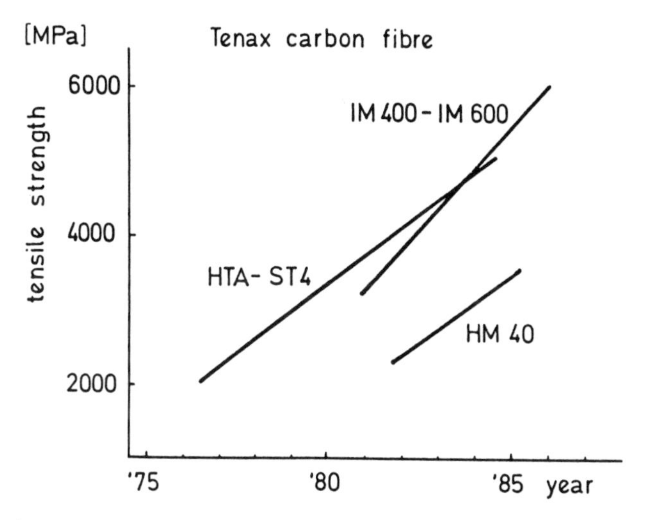

Figure 2 Relationship between tensile strength of carbon fibre and developing period

Properties of Twaron HM aramid fibre

This chapter will give a brief outline of the major properties of the Twaron HM aramid fibre as compared with the Tenax carbon fibre. This comparison will, however, be confined to just a few aspects as the general properties of aramid fibres have been amply described in the relevant publications (7 to 14). The properties to be dealt with will include those which have been gaining increasing importance because the corresponding field of application for aramids has become more significant or a more critical approach to the material has become necessary or because of major changes in test conditions. A comparison between the specific, i.e. mass-related, tensile moduli of elasticity and the tensile strength of aramid and carbon fibres is given in Fig. 3 showing a two-dimensional representation of the minimum and maximum fibre values under the same or analogous conditions (15).

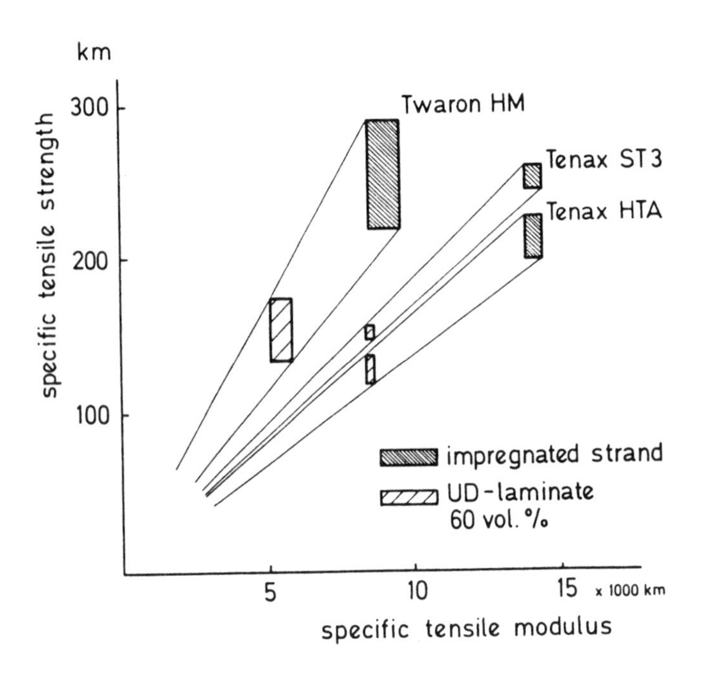

Figure 3 Specific tensile strength vs. specific tensile modulus of the aramid fibre TWARON HM and carbon fibre TENAX - types

From this graph it becomes evident that the new carbon fibres have come much closer to aramid fibres in terms of strength, although aramid fibres still exhibit the highest specific tensile strength.

The tensile modulus of elasticity and the methods of its determination have been amply discussed in recent years. Neither carbon nor aramid fibres have a constant tensile modulus over the whole elongation range. The graph shows the tensile modulus plotted against the elongation of the fibre subjected to tensile stress.

Figure 4 demonstrates that neither the modulus of the aramid nor the carbon fibre follows a linear course, with the individual carbon fibres behaving differently, while the modulus of Twaron HM is typical of all types of high strength aramid fibres. The carbon fibre selected was Tenax ST 3 which shows an almost steady increase of the tensile modulus. The modulus of aramid HM reaches its peak between 0.8 and 1.8 % elongation. One of the major reasons why aramid is selected for applications involving different speed, i.e. low (static) and requirement medium (aircraft) as well as very high velocities (ballistics), is its energy absorption. This high energy absorption of aramid, combined with its tough, laminar breaking behaviour and the non-harmful types of failure of aramid composites, are the main arguments in favour of this material.

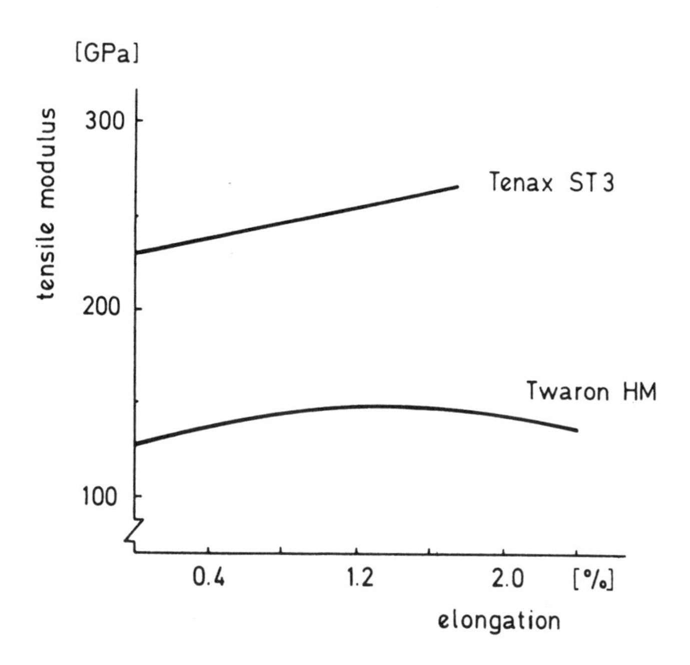

Figure 4 Differentiated curves of tensile modulus and elongation of aramid fibre TWARON HM and Carbon fibre TENAX ST 3

Figure 5 shows the almost static specific energy absorption of regular Twaron and Twaron HM as well as the carbon fibres Tenax HTA, ST 3, IM 500 and IM 600. The energy absorption values shown have been calculated on the basis of mean values of production data from the year 1984.

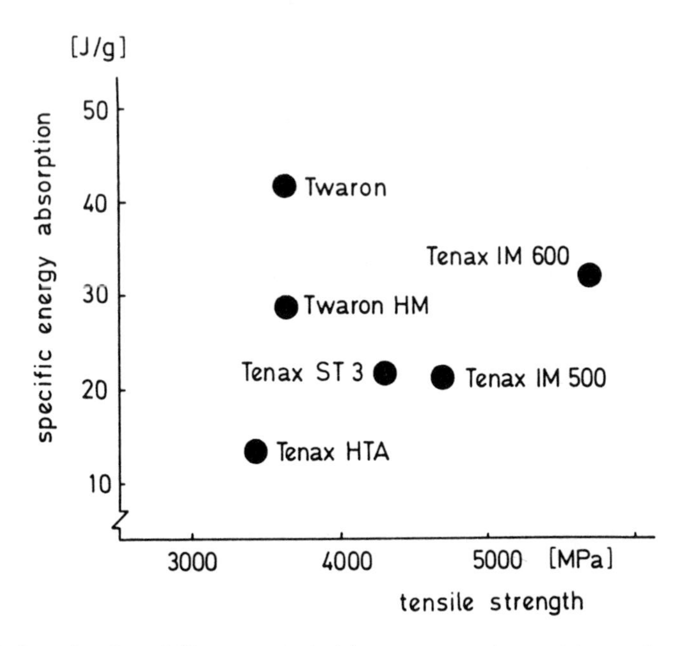

Figure 5 Specific quasistatic energy absorption of aramid fibre TWARON and carbon fibre TENAX

The carbon fibres IM 500 and IM 600 are the latest additions to this product range. This figure, where energy absorption has been plotted against tensile strength, clearly demonstrates the superiority of aramids. From this relationship we can derive a large number of applications. In the field of carbon fibres, only the most recent developments have come close to the potential of aramids.

Another property gaining increasing importance - as demonstrated by the use of aramid in primary structures, e.g. in aircraft construction, - is the resorption of moisture. Unlike carbon fibres, aramid does absorb moisture, and even rather quickly (Fig. 6).

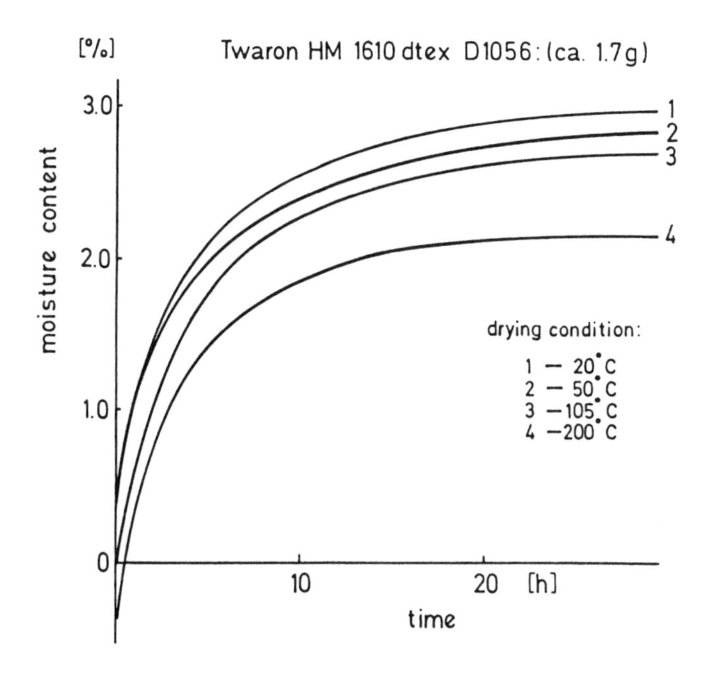

Figure 6 Water absorption after drying under different conditions

Resorption speed and moisture content vary with the temperature of the previous drying process. The dryness or residual moisture of the aramid fibres (desorption) depend on the drying temperature used (Fig. 7).

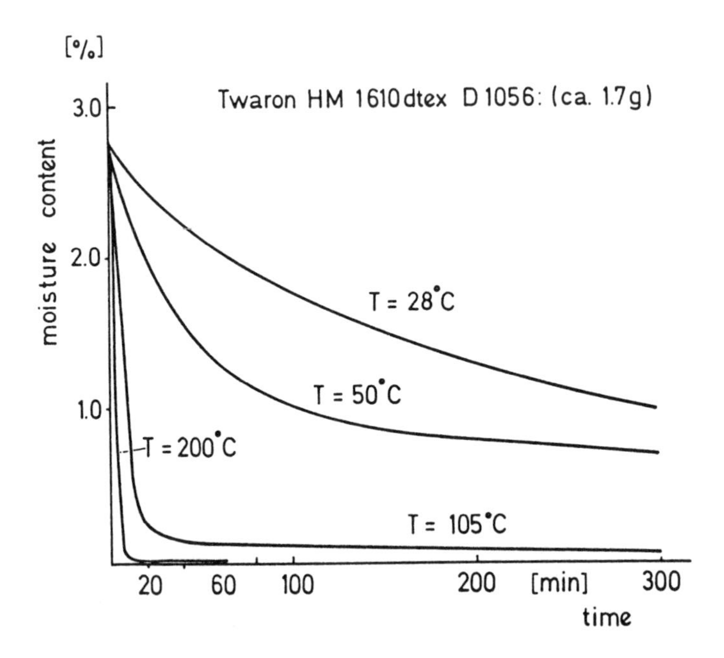

Figure 7 Desorption of Twaron HM under different drying conditions

A completely dry stage will only be reached at 105°C or higher temperatures, the time varying with the temperature used. Investigations on composites with aramid fibres of different moisture content have shown that the maximum interlaminar shear strength values as determined by the short beam test will - as a rule - be reached at a moisture content of less than 0.5 % (Fig. 8).

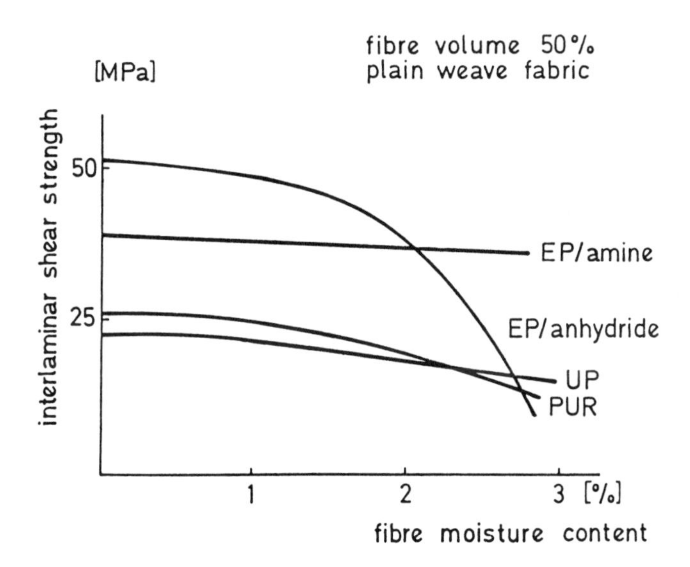

Figure 8 Moisture pick up of the aramid fibre related to short beam interlaminar shear strength

This means that the moisture content of aramid fibres should be lowered before processing to less than 1 %, preferably less than 0.5 %. The effect of moisture resorption on the other durability composite characteristics of aramid laminates have not yet been studied in detail. We are currently working on a joint test programme with the DFVLR (German Aerospace Research and Test Institutes). Another important factor in this type of application involving different temperatures and moisture levels, is the heat expansion coefficient.

Applications of TWARON aramid fibre In glider constructions, aramid-reinforced plastics are used for the fuselage as well as fairing elements. Aramid is normally used in the form of woven fabrics, in some cases also in carbon/aramid hybrid fabrics. Investigations on aramid-reinforced spars for gliders are currently being conducted by the DFVLR (German Aerospace Research and Test Institutes) with a view to obtaining general acceptance of aramid composites in glider construction. In aircraft construction in general, aramids are used in a wide variety of applications, ranging from cabin components of helicopters to doors and engine covers. In regular aeroplanes, aramid-reinforced plastics are found in main landing gear fairings, main landing gear doors, flap drag fairings and wing fuselage fairings. Typical examples are the F 50 of Fokker, SF 340 of Saab Scania and the DO 228 of Dornier. Interior aircraft components reinforced with aramid include kitchen parts, partitions walls, supply ducts and pilot seats.

In the industrial field, aramid-fibre-reinforced plastics are used in tubes and rods of high-voltage switch gears. Pressure tanks of aramid reinforced plastics are used for many purposes, e.g. as pressure energy accumulators, containers for fire extinguishing agents.

Another new field of application has been opened up by aluminium aramid-laminate composites (Fig. 9).

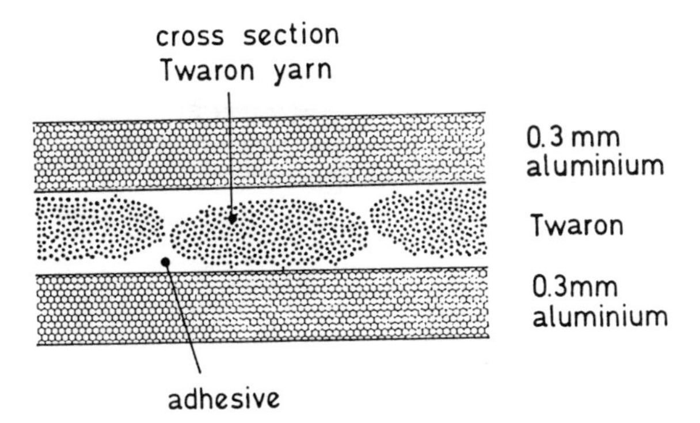

Figure 9 Aramid fibre aluminium sheet laminate

A major argument in favour of this composite, which is currently being tested in the aircraft industry, is the reduction of crack propagation below the level of monolithic aluminium. Fig. 10 gives a comparison between monolithic aluminium, type 2024-T3, and "Arall". These investigations have been carried out at the Instituut of Lucht- en Ruimtevaart of the Delft University of Technology. The other results obtained so far have also been judged very favourably, so that components will now be studied on a flying aeroplane.

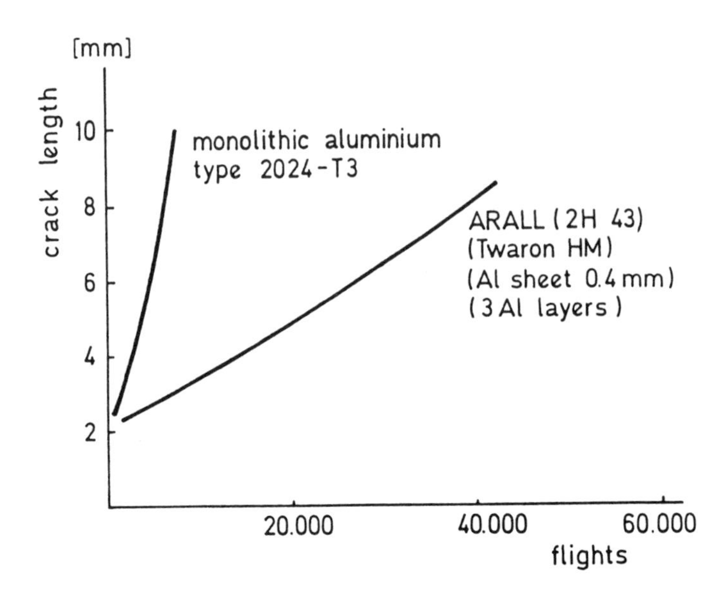

Figure 10 Crack propagation of aramid fibre aluminium sheet laminate

So far aramid fibres have predominantly been used for the reinforcement of thermosetting plastics but rarely thermoplastics. Owing to their specific chemical structure, thermosetting plastics are characterized by a high modulus of elasticity as well as excellent resistance to solvents and show low tendency to creeping even at elevated temperatures. But this also marks the limits to the use of thermosets, which are encountered in the mechanical properties of thermoset components as well as in the production of mouldings. The high modulus of elasticity is associated with low elongation at break as well as low energy absorption in failure and low toughness. The efficiency of the production of thermoset components is limited by the speed of the chemical setting reaction, which means that cycle times are longer than for thermoplastics.
In comparison with thermoset prepregs, thermoplastic semi-finished goods offer the advantage of unlimited storage life. Unlike thermoplastic parts, thermoset prepregs or finished parts cannot be shaped or welded.

Potential uses of aramid composites are vehicle components, such as shock absorbers, doors, large-size covering panels, ballistic laminate fairings for aeroplanes and industrial applications. Development work in this field is still in its infancy, but is increasingly gaining importance. Current objectives are the solution of technological production and processing problems as well as the development of new fibre-matrix adhesive systems, especially for thermoplastic composites.

In rocket technology aramid is used for fairings, e.g. in Ariane, but also for apogee engines and solid rocket engines for civil and military applications. In this sector it is again the high strength, high energy absorption and low density of the aramid fibre that make aramid/epoxide composites the material of choice. A relatively new application for composites is the protection against ballistic impact. The ballistic resistance of aramid fabric laminates is shown in table 1; the fibre content is about 70 % by volume. The fabrics are made of Twaron HM 121 tex. The resin used is polyurethane.

Fabric areal weight [kg/m]			3	6	9	18	36	54
Laminate thickness [mm]			3.3	6.6	10	20	40	60
Bullet	Velocity [m/s]	Energy [J]						
.357 Magnum KTW (10 m)	550	1100	●	●	●	●	○	—
.44 Magnum VMF (10 m)	450	1500	●	●	○	—	—	—
9mm Parabellum A1 B1 (10m)	420	700	●	●	○	—	—	—
7.65 mm Browning (10 m)	290	200	○	—	—	—	—	—

● complete penetration ○ cases of complete penetration
— no penetration

Table 1 Ballistic resistance of TWARON laminates

Fig. 11 shows the results of fragment simulating tests according to US Military Standard 662. Aramid laminates offer protection against both fragmention and bullets. For specific types of impact additional improvements can often be achieved by combination with other materials, e.g. ceramics or special steel alloys.

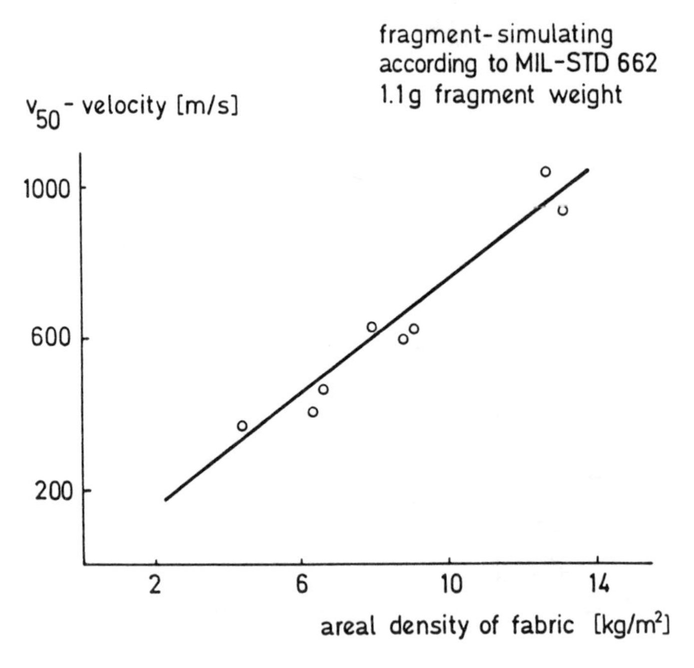

Figure 11 Ballistic test on TWARON aramid fibre laminate

A wide variety of possible applications exists, but there is still ample scope for development, especially in Europe. This applies to helmets for general and special military purposes, personal armour, rostrums, protection of property, armour for objects flying at low speed, such as helicopters or drones, vehicle armour, boat hulls for mine sweepers, bulkhead partitions for battle ships, superstructures of ships to protect persons and equipment, while simultaneously lowering the ship's centre of gravity and thereby improving its manoeuvrability. Fig. 12 shows a photograph of a helmet made of Twaron fabric whose weight-related ballistic resistance is comparable with high-performance steel alloys, but which offers still higher resistance to fragmentation.

Figure 12 Military helmet made of TWARON aramid fibre fabric laminate

SUMMARY

If we try to sum up again the major properties of aramid fibres from which the multiplicity of composite uses derives, we will have to name their extremely high tensile strength, low density, high energy absorption and their "good-natured" laminar breaking behaviour. In view of their limited compressive strength, aramid fibre composites are predominantly used in components subjected to tensile stress or high-performance coverings.
The picture is rounded off by good dielectric properties, X-ray transparency, heat and cold resistance and resistance to chemicals as well as good abrasive properties.

REFERENCES

1 Hillermeier K, Kunststoffe (Plastics) 63 (1973) 899

2 Celanese GB, Patent 979.342 (1961)

3 Monsanto US, Patent 3.414.645 (1964)

4 Akzo US, Patent 4.308.374 (1975)

5 Teijin, 'Technical Information on HM-50' (1982)

6 Zahr G E, Sampe Conference, Scheveningen (1985)

7 Hillermeier K, Kunststoffe (Plastics) 66 (1976) 802

8 Hillermeier K, Plastica 30 (1977) 374

9 Hillermeier K, VDI Edition (1977) 29

10 Northolt M G, Polymer 21 (1980) 1199

11 Blumberg H, Hillermeier K, Krueger R, Melliand 63 (1981) 7

12 Hupjè W, Enka MRG Symposium (1982)

13 Blumberg H, Estratto da plast 12 (1982)

14 Hillermeier K, Textil Research Journal 54 (1984)

15 Enka, 'Testmethode for strand test' (1984)

16 Marissen R, Vogelsang L B, Sampe Conference, Cannes (1981)

17 Stolze R, Milano Plast '85 (1985)

Karlheinz Hillermeier studied fibre technology and engineering. Since the early seventies he has been employed with Enka Ag. He is engaged in the fibre and application development of high modulus fibre since then, and is head of the material development and technical service for composites.

HIGH MODULUS FIBERS AND THEIR COMPOSITES

N.H. Ladizesky and I.M. Ward

Department of Physics
University of Leeds

In this paper several aspects of recent research at Leeds University on ultra high modulus polyethylene (UHMPE) fibres and other composites are described. Comparative data are presented for unidirectional fibre composites where the fibre phase is either wholly UHMPE, Kevlar 49, Carbon EXAS or E-type glass. Results are also presented for hybrid composites which incorporate both UHMPE fibre and one of the commercial fibres. It is shown that these hybrid composites can combine in a useful way the good qualities of UHMPE fibres, especially their very high energy absorption, with other good qualities of the commercial fibres.

INTRODUCTION

The research described in this paper stems from an ongoing research programme at Leeds University on ultra high modulus polyethylene (UHMPE) fibres. There are several major aspects to this research programme. First, there is the development of practical processes for the production of UHMPE fibres by the melt spinning and drawing route. Basic research on the tensile drawing of linear polyethylene in the early 1970s led to the setting up of a small pilot plant process at Leeds, for the production of monofilaments and multifilament yarns at economic rates (Capaccio et al (1-3)). The transfer of this technology under licence to the Celanese Fibers Company USA, has led to the availability of fibres for large scale development applications trials. Secondly, there is the continuing quest for improved fibre properties, especially with regard to creep behaviour, improved temperature stability and ultimate strength (Wilding and Ward (4)). This has led to some notable recent developments, for example in developing cross-linking treatments (Woods, Busfield and Ward (5)). Finally, there is the exploration of many areas of potential application for these fibres, including the reinforcement of brittle matrices, of which polymer resins form a major part. In this paper the development of UHMPE resin composites will be described in the context of our present development of the UHMPE fibres and in comparison with the performance of composites reinforced with presently available commercial fibres.

ULTRA HIGH MODULUS POLYETHYLENE FIBRES

Fibre properties

Table 1 shows the tensile properties of melt spun and drawn UHMPE fibres,

TABLE 1. Properties of reinforcing fibres (room temperature)

Property Fibre	Tensile Modulus (GPa)	Tensile Strength (GPa)	Elongation at Break %	Density ρ g/cm^3	Specific Modulus GPa/ρ	Specific Strength GPa/ρ	Maximum Working Temperature °C
Carbon	250	3.6	1.5	1.80	139	2.0	>1500
Glass	75	3.0	2.5	2.54	30	1.2	250
Kevlar 49	125	3.0	3.0	1.45	85	2.1	~ 180
Polyethylene	40-70*	1-1.5	4-18**	0.96	42-73*	1-1.5	130

* Depending on draw ratio.
** Depending on strain rate.

together with those for carbon, glass and Kevlar fibres. The best available properties are shown for the commercially available fibres, whereas those for UHMPE fibres are conservative, based on properties which can readily be achieved by the low cost melt spinning and drawing route. Appreciably higher strength fibres can be obtained by the gel spinning and drawing route (Smith and Lemstra (6)), which is an intrinsically more costly process.

The most appropriate comparison can be made on the basis of the specific properties, which takes into account major differences in density. On this basis, UHMPE fibres show properties which challenge those of glass and Kevlar fibres, and although their stiffness is much less than that of carbon fibres, they show a very high breaking extension. This suggests that the total energy to break will be high, and it will be shown that this feature can be of considerable importance in determining the properties of UHMPE composites.

Recent developments in UHMPE fibres

It was recognized from the outset that UHMPE fibres might be of restricted application, in spite of the very high tensile properties as shown in Table 1, because of creep deformation when subjected to continuous loading and a limited temperature range. Although improvements can be made by processing high molecular weight polymer (Wilding and Ward (4)), notably for the costly gel spinning route, even for these fibres the high initial strengths of ~ 2.5 GPa are reduced to constant loading strengths of ~ 0.5 GPa. In the case of melt spun and drawn fibres, it has been shown that the creep resistance can be dramatically improved by cross-linking procedures and guidelines have been established for electron beam irradiation of bulk fibres. One successful treatment method involves exposing fibres to a radiation dose of 20 MRad in acetylene (Woods, Busfield and Ward (5)). Creep is then virtually eliminated, so that the fibres can be subjected to cyclic strains for very long periods without permanent flow additionally, the tensile properties are retained up to temperatures of 130°C for long periods of time, and even up to higher temperatures for the short periods of time required for some processing procedures.

For ballistic applications and impact performance, the properties at high strain rates are of primary importance. Detailed studies at Leeds show that tensile strengths in the range 1.5-2.0 GPa can be achieved with melt spun and drawn fibres (Cansfield et al (7)). Although the strength depends primarily on the number average molecular weight $\bar{M}_n$, higher strengths are also obtained for higher weight average molecular weights $\bar{M}_w$ at a given level of $\bar{M}_n$, and there does seem to be scope for making very satisfactorily products from a wide range of grades of polymer (Hallam (8)).

UHMPE FIBRE COMPOSITES

Fibre-resin adhesion

It was appreciated that the chemical inertness of polyethylene, although an advantage in many respects would militate against a good bond between UHMPE fibres and polymer matrices. A comprehensive study of fibre-resin adhesion was therefore undertaken (Ladizesky and Ward (9)), and for convenience the preliminary research was conducted using monofilaments ~ 0.4 mm diameter. The adhesive strength was determined by embedding one end of a monofilament in a disc of polymer resin, most of the work being undertaken using a low viscosity resin intended for high strength composites (Ciby Geigy epoxy resin XD927). Two principal surface treatments were examined

(a) Immersion in acid solutions, primarily chromic acid
(b) Plasma etching in the presence of a carrier gas.

Although both acid and plasma treatments were effective in raising the pull-out adhesion for low draw ratio monofilaments (draw ratios 8 and 15), for the high draw ratio monofilaments (draw ratio 30) the optimum treatment was plasma etching in the presence of oxygen. In the latter case pull-out adhesion values of about ten times those for untreated monofilaments were obtained. Scanning electron microscopy showed that the plasma etching produces a fine cellular structure on the surface of the monofilaments. The resin penetrates this structure to give a mechanical bond between fibre and resin. The effectiveness of the plasma treatment does depend partly, however, on the improvements in wettability of the polyethylene fibre by the resin prior to curing.

It was shown that very severe surface treatment, either by acid or plasma etching does reduce the tensile strength of the monofilaments. In extending and developing the plasma etching method for the surface treatment of multifilament yarn, conditions were established where the cellular structure could be produced and give excellent improvements in fibre-resin adhesion with undetectable reduction in yarn strength, as will be apparent from the mechanical properties of the composites.

Preparation of fibre composites

Fibre composites were prepared by three methods.

(a) The leaky mould technique In this technique a bundle of fibres is placed in a rectangular mould, fully wetted with liquid resin, and then compressed with a smooth fitting top to the mould. Excess resin is squeezed out of the mould and the system cured. This technique produces a rectangular bar with the fibre orientation along the length of the bar.

(b) The wet lay-up technique This technique is similar to the leaky mould technique, but layers of square weave fabric are laid down between layers of liquid resin in an identical rectangular mould, instead of bundles of fibres. The warp and weft of the fabric are parallel to the principal axes of the rectangular mould.

(c) Lamination of pre-impregnated sheets of fibre (pre-pregs) This is the commercial route for fibre laminates. Pre-pregs of UHMPE, carbon, glass and Kevlar fibres were prepared by Rotorways, Bridgewater UK using Code 91 epoxy resin. The pre-pregs were moulded in a hot press to produce both homo fibre and hybrid fibre laminates.

In all cases the fibre volume fraction was 55%.

Mechanical tests on composites

The mechanical tests on the composites followed standard procedures established at the Royal Aircraft Establishment (RAE), Farnborough.

Three point bend tests using specially constructed test rigs, were used to determine the flexural modulus (FM), ultimate flexural strength (UFS) and inter laminar shear strength (ILLS). In all cases rectangular specimens were prepared from the fibre composites, the dimension being varied appropriately, as described in ref 10. The FM was determined for an equivalent tensile strain of 0.03%.

The measurements of tensile strength (TS), tensile modulus (TM) and compressive strength (CS) required the preparation of samples whose ends were sandwiched between soft aluminium alloy plates bonded to the sample surfaces. The gauge length varied from 50 mm for the TM measurements to 10 mm for CS measurements. For the TS measurements the sample was waisted with a continuous radius of 1000 mm, giving a minimum thickness of ~ 1.2 mm. Full details of these tests have been given elsewhere (Ewins (11) Curtis and Martin (12)).

Charpy impact tests were undertaken on unnotched specimens of dimension 8 cm x 1 cm x 2 mm, impacted on the broad surface of the bar, again following procedures advocated by RAE Farnborough (Dorey (13)).

Experimental results on fibre composites

The key experimental results are summarized in Tables 2, 3 and 4, and in

TABLE 2. Mechanical properties of "leaky" mould composite systems - continuous UHMPE yarn/epoxy XD927 resin

Reinforcement Orientation	Reinforcement Treatment	ILSS MPa	UFS Test: Value MPa	UFS Test: Decrease After 3 Successive Tests (%)	FM GPa	TM GPa	TS GPa	CS MPa	"Flat" Charpy Test-Energy Absorption: At 1st Impact kJ/m^2	"Flat" Charpy Test-Energy Absorption: Decrease After 3 Successive Tests (%)
Unidirectional	Untreated	15	165	8	22	19	0.31	80	160	70
	Acid treated	20	145	7	—	20	0.33	83	—	—
	Plasma treated	27	150	5	19	21	0.33	85	120	85
Woven	Untreated	17	85	5	8	9	0.12	70	60	85
Reinforcement (0/90)	Plasma treated	24	95	4	9	9	0.13	80	45	80

1. All measurements at room temperature.
2. All composites about 55% reinforcement by volume.

Figures 1 and 2.

Leaky mould measurements were used primarily to examine the influence of surface treatment on ILLS values. It can be seen from Table 2 that the improvements in ILLS following surface treatment are in line with expectations based on the pull-out adhesion of monofilaments, but at a lower level of improvement. The ILLS of the composites increases by only a factor of about two for the optimum plasma etching treatment compared with an order of magnitude improvement for the monofilaments. The tensile properties of the UHMPE homofibre composites is roughly in line with expectations, although it will be seen by comparing Tables 2 and 3 that the commercial pre-preg route does give

TABLE 3. Mechanical properties of pre-preg composite systems - various continuous reinforcements/Code 91 epoxy resin

Reinforcement	Density kgm^{-3} (x10^{-3})	ILSS MPa	. . UFS Test . . Value MPa	Decrease After 3 Successive Tests (%)	FM . .	TS GPa	TM . . .	CS MPa	"Flat" Charpy Test-Energy . . Absorption . . At 1st Impact kJm^{-2}	Decrease After 3 Successive Tests (%)
UHMPE (untreated)	1.08	15	165	27	41	0.43	41	75	135	85
Kevlar 49	1.35	53	532	28	66	1.30	75	277	175	90
Carbon EXAS	1.56	66	1600	Break	104	1.95	137	1050	75	Break
E-Type Glass	1.95	66	1145	Break	41	1.56	56	975	320	Break
UHMPE (untreated)-Carbon EXAS	1.28	34	495	2	48	—	85	410	155	90
UHMPE (untreated)-E-Type Glass	1.45	28	245	22	42	—	46	255	240	30
UHMPE (untreated)-Kevlar 49	1.19	22	285	5	41	—	57	160	135	90

1. All measurements at room temperature.
2. All composites about 55% reinforcement by volume.
3. Hybrids: 2 x UHMPE fibers/3 x commercial fibers/2 x UHMPE fibers.

TABLE 4. Specific mechanical properties of pre-preg composite systems - various continuous reinforcement/Code 91 epoxy resin

Reinforcement	ILSS MPa/ρ	UFS MPa/ρ	FM GPa/ρ	TM GPa/ρ (x 10^3)	TS GPa/ρ	CS MPa/ρ	"Flat" Charpy Test-Energy Absorption at 1st Impact kJm^{-2}/ρ (x 10^3)
UHMPE (untreated)	14	155	38	38	0.40	70	125
Kevlar 49	39	394	49	56	0.96	205	130
Carbon EXAS	42	1025	67	88	1.25	675	50
E-Type Glass	34	585	21	29	0.80	500	160
UHMPE (untreated)-Carbon EXAS	27	385	38	66	—	320	120
UHMPE (untreated)-E-Type Glass	19	170	29	32	—	175	165
UHMPE (untreated)-Kevlar 49	18	240	34	48	—	135	115

ρ : Density as given in Table 1.

better results. The compressive strength of the composites is low, reflecting the low compressive strengths of UHMPE fibres, in common with other organic fibres, such as Kevlar.

All the UHMPE homofibre composites show excellent impact performance, and moreover the composites do not shatter, but retain their integrity after impact, in contrast to carbon and glass fibre composites. This also holds for the UHMPE/carbon and the UHMPE/glass fibre composites. Moreover, these hybrid composites enable the advantageous tensile properties of carbon and glass to be utilized. For example the UHMPE/carbon fibre composite has comparatively high UFS and CS.

The advantages of the hybrid composites incorporating UHMPE fibres are further exemplified by the comparison of specific properties shown in Table 4, and by the visual evidence for their ductility illustrated by the comparisons of Figures 1 and 2.

CONCLUSIONS

It has been shown that the unique combination of fibre properties possessed by UHMPE fibres can be used to advantage in fibre composites. From the viewpoint of the present conference on carbon fibres, the results obtained for hybrid UHMPE/carbon fibre composites are of most interest. In these composites the high energy absorption capabilities of the UHMPE fibres can be combined with the very high stiffness and strength of carbon fibres to produce a composite material of outstanding all-round properties.

ACKNOWLEDGEMENT

We wish to thank Mr. L.N.Phillips, Mr. W. Johnson and Dr. J. Harvey of RAE Farnborough for useful discussions and invaluable support during the course of this research.

REFERENCES

1 Capaccio G and Ward I M , Nature, Phys. Sci 243 (1973) 143

2 Capaccio G, Crompton T A and Ward I M J Polymer Sci., Polymer Phys Edn 14 1641; 18 (1980) 301

3 Capaccio G, Smith F S and Ward I M, British Patent No. 1506565 (filed 5 March 1974)

4 Wilding M A and Ward I M Polymer 19 (1978) 969; 22 (1981) 870

5 Woods D W, Busfield W K and Ward I M Polym Comm 25 (1984) 298

6 Smith P and Lemstra P J, J Mater Sci 15 (1980) 505

7 Cansfield D L M, Ward I M, Woods D W, Buckley A, Pierce J M and Wesley J L, Polym Comm 24 (1983) 130

8 Hallam M (unpublished work)

9 Ladizesky N H and Ward I M J Mater Sci 18 (1983) 533

10 Sturgeon J B, Technical Report 71026, Royal Aircraft Establishment, Farnborough UK (1971)

11 Ewins P D, Technical Report 71217, Royal Aircraft Establishment, Farnborough UK (1971)

12 Curtis P T and Martin J, Progress in Science and Engineering of Composites, The Japan Society for Composite Materials 1 (1982) 219

13 Dorey G, Royal Aircraft Establishment, Farnborough UK, Personel Communication.

Dr N H Ladizesky obtained a "Licenciado" in Physical Sciences from the National University of Cuyo, Argentina in 1958. In 1960 he went to Australia to CSIRO as an Experimental Officer for one year followed by three years with ICI Fibres Ltd, UK, as a Technical Officer. He joined Bristol University in 1967 as Research Assistant, moving to Leeds University in 1969. During 1972 he gained a PhD from Bristol University in Polymer Physics. After spending three years in Malaysia as a Lecturer (1973-76) he became a Research Fellow at University of Glasgow before returning to Leeds University in 1979 where he is now a Senior Research Fellow.

Professor I M Ward read Physics at Oxford, and subsequently undertook postgraduate research on the electron spin resonance of inorganic complexes and irradiated materials. From Oxford he joined ICI Fibres Division, where his special responsibilities included infra-red spectroscopy; nuclear magnetic resonance and physical properties of fibres. In 1982 Professor Ward was appointed Head of the Basic Physic Section, becoming an I.C.I. Research Associate in January, 1965. Returning to academic life in 1966 as Senior Lecturer in the Physics of Materials at the H.H. Wills Physics Laboratory University of Bristol he was appointed to Chair in Physics at the University of Leeds in 1970. The Polymer Group which he founded at Leeds is concerned primarily with the structure and properties of solid polymers with particular interest in the development of very highly oriented polymers. In 1983 Professor Ward was elected a Fellow of the Royal Society.

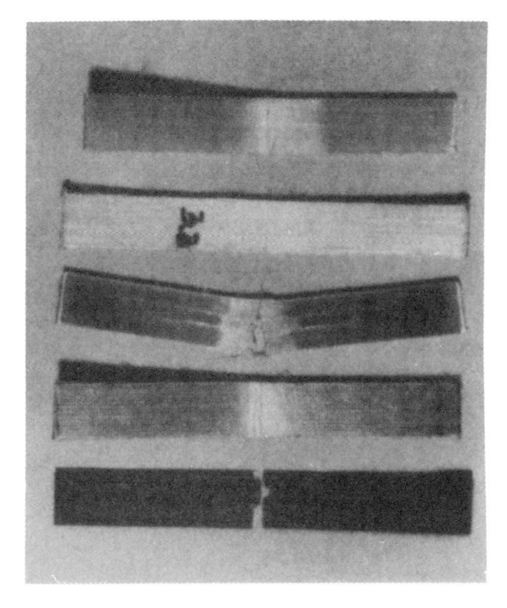

Figure 1 Unidirectional composites after "flat" Charpy test.
a) UHMPE - 7 impacts b) UHMPE/E-Type Glass - 7 impacts c) E-Type Glass - 1 impact
d) UHMPE/Carbon EXAS - 7 impacts e) Carbon EXAS - 1 impact

Figure 2 Unidirectional composites after UFS test - 0° samples.
a) UHMPE - 3 cycles b) UHMPE/E-Type Glass - 3 cycles c) E-Type Glass - 1 cycle
d) UHMPE/Carbon EXAS - 3 cycles e) Carbon EXAS - 1 cycle

HYBRID SYSTEMS—BEST OF BOTH WORLDS?

J.V. Weaver and E. Taylor

Fothergill Engineered Fabrics Ltd.
Littleborough, Lancs.

Hybrid systems - defined within the scope of this paper as two or more materials in fibrous form used in the same matrix, can significantly expand the range of properties which can be achieved in reinforced composites; at the same time, they can offer a way of controlling costs which might otherwise preclude the use of some fibres.

INTRODUCTION

Generally, fibre reinforced composites are regarded as one means of providing materials which exhibit high specific strength and stiffness. Accordingly, they have long enjoyed considerable success in a substitutional role for some of the more conventional materials such as steel, and even wood. However, the successive development of new fibres and the coincident improvements in existing fibres, together with new and improved matrices, has meant that modern composites can replace not only "conventional" materials, but also earlier composites which perhaps have come to represent the accepted standard in some applications.

Material selection for specific applications depends upon a number of factors, each of which can be considered as a constraint on the suitability of candidate materials for the job in hand. Whilst the checklist of constraints, and certainly the order of priority for individual parameters within the specification, will vary from case to case, typically consideration will be given to the physical, mechanical and often chemical characteristic requirements, and to availability, processability, and cost. Arguably, the latter is the greatest driving force, the key factor for selection.

In simple terms, therefore, there are two sides to the selection analysis. Firstly, definition of the material property requirements; secondly, definition of the constraints on the more general material attributes, such as cost. In terms of hybrid reinforcement systems, it is possible to address both of these elements, literally engineering reinforcements to take advantage of particular fibre material properties, whilst introducing a level of cost control which may not otherwise be achieved.

HYBRID REINFORCEMENT SYSTEMS

There are a number of ways in which different fibres or materials can be combined to reinforce the same matrix. Fig. 1 shows a simplified summary, indicating the three basic levels at which material combination can be achieved.

On the smallest scale, "composite" filaments are available, typical examples being coated fibres such as aluminium on glass, nickel-plated carbon, and silicon carbide on boron. Such fibres represent an interesting group of materials, since the coatings themselves are modifying the base fibre properties in terms not just of the mechanical function, but also electrical and thermal characteristics, and for BORSIC, in terms of the fibre chemical compatibility with metal matrices at elevated temperature.

With yarns, the benefit of combining different fibres may be most apparent in terms of the handling of some materials, for example in the production of reinforcement fabrics. Hybrid yarns can be particularly important in facilitating the processing of brittle fibres.

At the highest level, hybrid fabrics are probably the most widely recognised form of multimaterial reinforcement usually in woven or unidirectional tape versions. Material combination within the matrix can be achieved in two ways - using two or more fibres in the same fabric (the true intraply hybrid), or by mixing single fibre plies through the composite cross-section - referred to as interply hybridisation. Both types are shown in Fig. 2.

Of course it is possible to create quite complex hybrids, for example by using a hybrid filament to produce a hybrid yarn and thence a hybrid intraply fabric. A typical case is a metallised glass fibre/Kevlar fibre hybrid which can be used to fabricate electrically conductive, lightweight, stiff composites as used, for example in some microwave and lightning strike protection applications.

DESIGNING COMPOSITE PROPERTIES

Property requirements for composite materials are becoming increasingly complex. In many instances, it is no longer sufficient to describe a simple material profile of strong and stiff at the lowest possible weight. Mechanical, physical and chemical environments in which composites are required to operate are frequently becoming more hostile. In consequence, not only the level, but also the combination of composite properties can be more severe and complex.

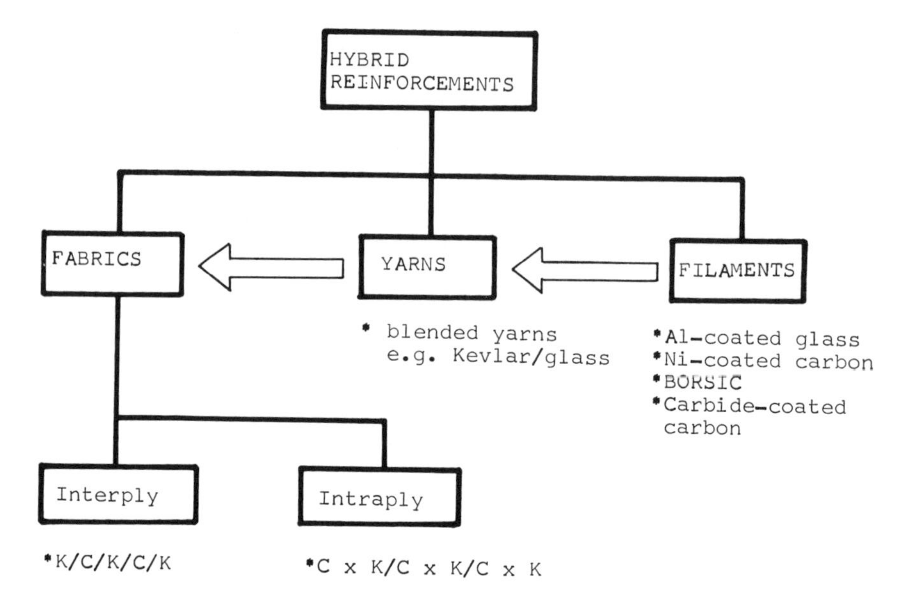

Figure 1 Levels of Hybridisation

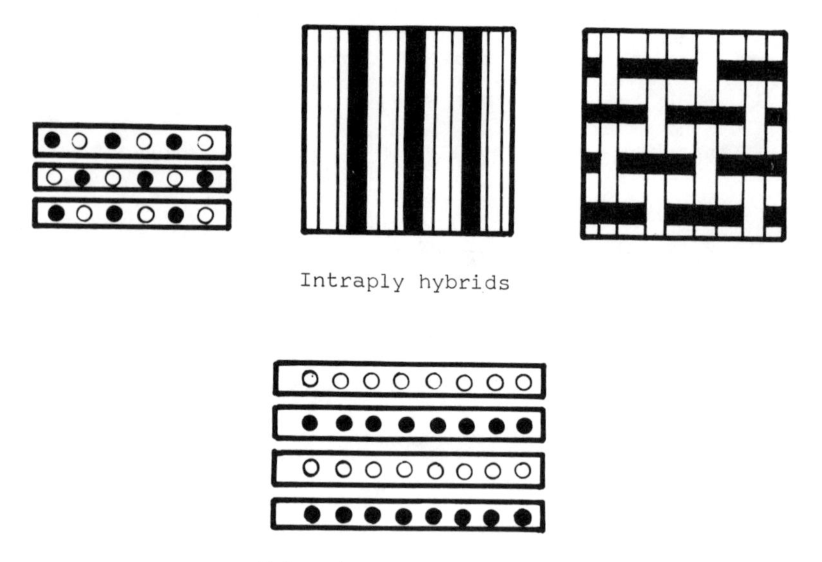

Figure 2 Hybrid fabric reinforcement

If we consider a composite consisting of a matrix and a particular reinforcing fibre, then in general, composite properties will depend upon the amount of fibre, hence matrix, which is present. If, as is the usual case, the fibre is both stiffer and stronger than the matrix, then composite stiffness and strength will increase with increasing fibre volume. However, there are practical constraints on the maximum loading of fibre which can be achieved, depending upon the actual form of the reinforcement, as shown in Fig. 3, for glass/polyester, Furthermore the degree of material anisotropy changes with the form of reinforcement, which itself may affect design considerations.

The range of alternative reinforcement fibres presently available provides a further opportunity for modifying composite properties. However, it is increasingly important to consider more general composite behaviour rather than a single property, such as stiffness in isolation. For example, one area of concern, particularly in structural applications, is that of response to impact loadings, or damage tolerance. Unlike metals, which can deform by plastic flow, most organic matrix composites can only absorb energy elastically, or by various fracture processes. Energy absorption during loading is related to the area under the stress-strain curve, proportional to σ^2/E. Typical curves for common reinforcements are shown in Fig 4. High modulus fibres such as carbon and boron require relatively little energy to reach failure, at strains of the order of 1-1.5% (though more recent carbon fibres claim around 2% failure strain). Composite

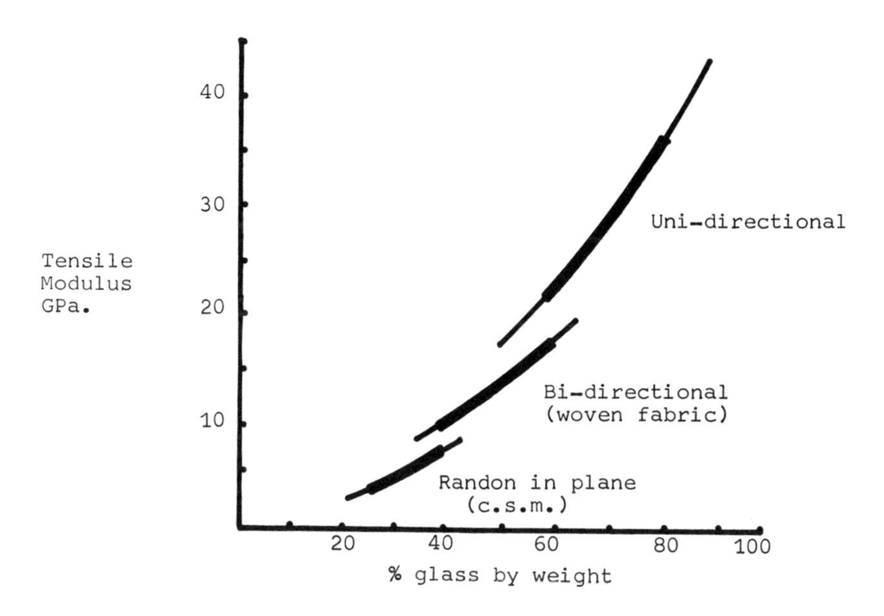

Figure 3. Typical fibre loadings

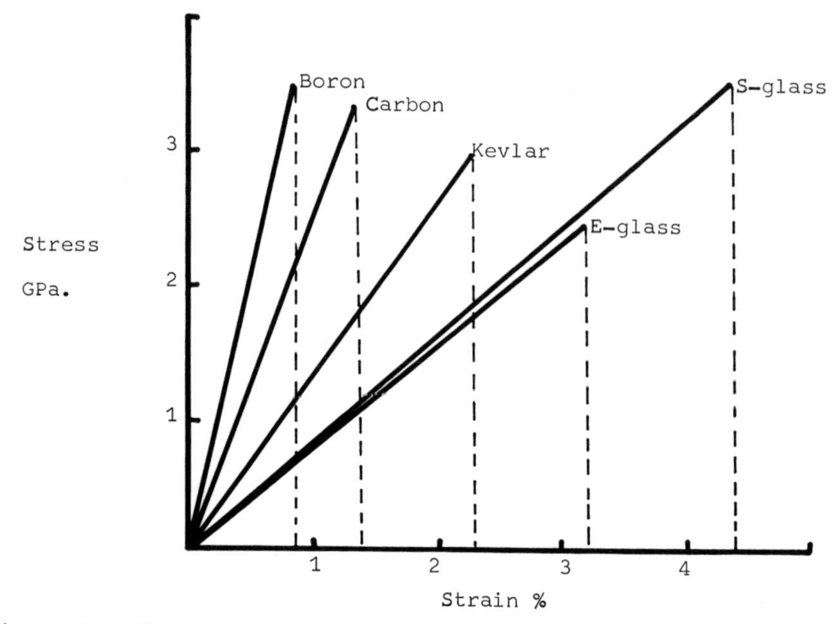

Figure 4. Stress-strain curves for common reinforcement fibres

fracture is characteristically brittle in nature. Fibres such as Kevlar and glass, whilst failing at similar stresses, do so at higher strain levels, signifying higher energy absorption. It is reasonable to expect therefore that combining different fibres in the matrix - hybridisation - offers a third 'degree of freedom' in engineering a balance in composite properties, although balance will always imply some level of compromise. Typical experimental results illustrate this compromise, as in Fig. 5 presented by Dorey et al (1) for carbon/Kevlar interply epoxy resin composites. Although impact resistance is enhanced by the addition of Kevlar to the carbon fibre composite, properties such as modulus, flexural and compressive strength are attenuated. Better damage tolerance, in terms of residual strength after impact was reported in these interply composites when the stacking sequence involved Kevlar, rather than carbon, outer plies, indicating a shielding of the 'core' carbon fibre plies.

Much of the early work on hybrids has involved the development of model structures useful in the study of composite behaviour. Carbon/glass hybrids with relatively large fibre property differences, have received particular attention. A number of comprehensive reviews have been presented (2-5) wherein a common theme is the reported observation of a "hybrid effect" in such composites. Different interpretations are apparent, but the basic observation is that low elongation fibres, like carbon, fail at higher strain levels when in combination with more extensible fibres such as glass, than when used as the sole reinforcement.

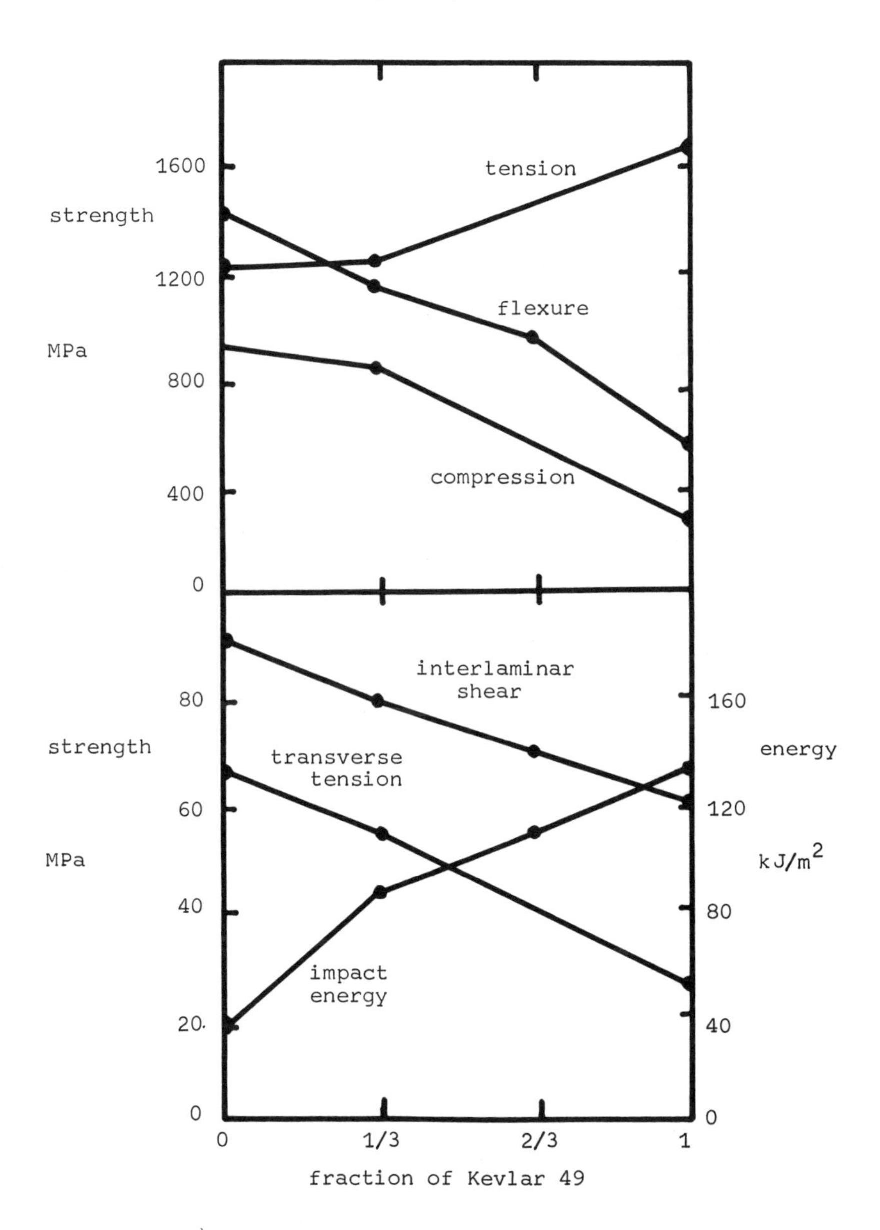

Figure 5. Carbon/Kevlar - epoxy composite properties. Vf = 60%; u/d interply hybrids. (Ref. 1)

Fig. 6 shows the theoretical strength of carbon/glass hybrid composites (5). Points A and D are the failure stresses for all-glass and all-carbon materials respectively. The line BD shows the stress at which the carbon fibre element would be expected to fail. Where the carbon fraction is sufficiently small (AC), failure in this phase can be supported by the glass fibres, which determine composite strength along the line AC. Beyond point C, at high carbon fractions, failure of the carbon (along CD) causes transfer of load which cannot be sustained by the glass, which also fails. Whilst tensile modulus can be predicted arithmetically by the Rule of Mixtures, dependant upon the relative proportions of fibres present, strength is not represented in the same manner, for example, along the line AD. Theoretically, the simple mixtures rule would be valid only in the special cases in which both fibres have the same failure strain, or where stress remains constant and is independent of strain (e.g. yielding or fibre pull-out).

The hybrid effect has been supported by observations where the initial failure stress (of the carbon) rises above the line BCD, or ultimate composite strength above the line ACD; such variations have been related to fibre ratio within the hybrid and the separation, or dispersion, of the higher stiffness fibre. In carbon/glass systems, the effect appears to be greater at finer dispersions and lower ratios of the stiffer, carbon fibre.

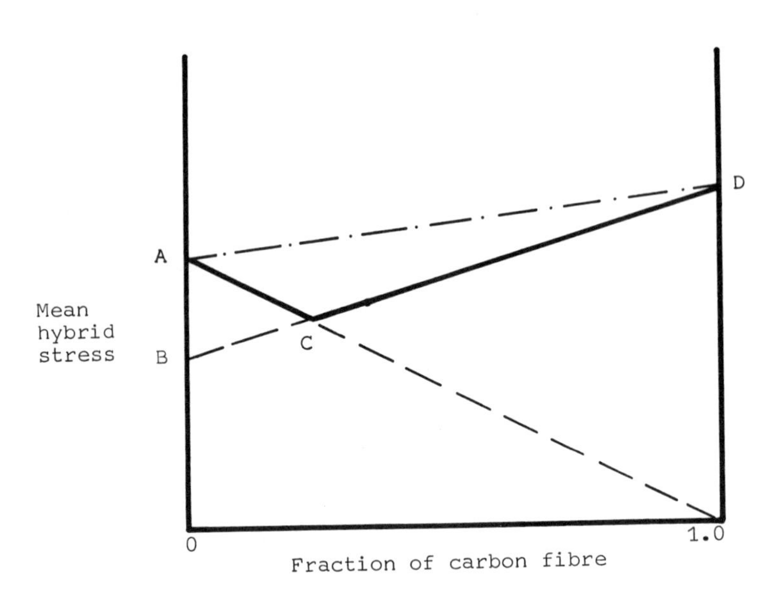

Figure 6. Theoretical strength of carbon/glass hybrid composites. (Ref. 5)

The hybrid effect debate merely illustrates an important concern over the application of hybrid reinforcements, that of predicting accurately the final composite behaviour. This, together with the acceptance of compromise inherent in such systems, implies specific hybrid design for particular applications; this may lead to extended test and evaluation programmes both for the reinforcement and final composite.

Weight Sensitivity

Since reduction in component weight is often a prime objective in using a composite structure, it is important to recognise that hybrid reinforcement itself can provide significant additional benefits in weight saving, specifically in cases where lower density fibres are introduced to the reinforcement, or where component thickness can be reduced through hybridisation. Hybrids thus can offer some balance between the requirements of weight and cost; of the common reinforcements, glass fibre is cheapest, but heaviest, implying a direct conflict in the cost-weight equation.

BEYOND THE MECHANICS

Material selection depends upon factors other than the basic demand for meeting intrinsic composite properties. The practical application of hybrid reinforcements has highlighted some of these other considerations.

Cost

In simple terms, hybridisation may afford cost benefits or penalties, depending upon whether more expensive or cheaper fibres are being replaced in the original reinforcement. Unfortunately, cost is often regarded in these simple material cost terms, whereas in reality, there are a number of elements which contribute to total cost. For example, in situations where fewer plies of hybrid reinforcement are dictated by the design considerations, there may be significant savings in fabrication times, resin use, etc. Such "knock-on" effects may outweigh the (more obvious) raw material cost penalties. Typically costs of carbon, Kevlar and glass reinforcements are currently in the ration 8:4:1.

Availability

The concept of designing specific hybrid constructions for individual applications, at least in terms of intraply fabrics, implies a successive move away from standard "styles" of construction which have been developed over the years. Intraply hybrids as yet are not truly considered as "stock" items.

Whilst the implication may be one of paying a premium for a hybrid reinforcement, this is not generally true in practice. Where hybrids involve more expensive fibres, such as carbon, it is often the case that shorter production runs are acceptable in manufacturing terms; accordingly, constraints on minimum quantity availability may be less stringent.

The argument is less apparent for interply hybridisation, since in many cases it is possible to design around more conventional styles of fabric (or tape) - a reason why emphasis has been given to this approach, despite the potential penalties of cost and handling associated with using two separate, rather than single, plies of material.

Fibre Finish - Coupling Agents

Most of the common reinforcements are treated with oil, or some form of size designed to aid handling. Such treatments tend to be fibre specific. In the case of woven fabrics, post-treatments are frequently used, since it is necessary to remove processing treatments and replace them by proprietary coupling agents or finishes to provide selective compatibility with various matrix systems. By and large, carbon fibre fabrics do not undergo such post-fabrication treatment. Conversely, aqueous or solvent-based scouring treatments may be used for aramid fabrics; glass fabrics are conventionally heat-treated prior to finishing. Hence, typical post-treatments may not be appropriate for intraply hybrids, particularly those containing carbon. Subsequent finishing with fibre specific agents again may not be suitable. In practice, however, these difficulties can be minimised by careful selection of fibre grades, and with proper attention to fabric finishing. Nonetheless some particular problems have yet to be resolved.

Composite Machining

Different machining techniques have been established for different single fibre composite materials; it is not surprising, therefore, that a number of difficulties have been apparent in the cutting, drilling and shaping of hybrid systems. Carbon and glass composites are relatively easy to machine, Kevlar rather less so. Significant improvements have been found when Kevlar is combined with carbon or glass, in intraply and interply hybrids. In the latter case it is better from a machining standpoint to have carbon or glass outer plies, though this conflicts with the greater improvement in fracture toughness afforded by Kevlar outer layers.

On balance, there would appear to greater machining benefits when using intraply fabrics, in view of some evidence of delamination in interply composites.

REFERENCES

1 Dorey G, Sidey G.R and Hutchings J, Composites 9 (1978) 25

2 Hardaker K.M and Richardson M.O.W. Polym Plast Tech Eng 15 (1980) 169

3 Short D and Summerscales J, Composites 10 (1979) 215

4 Short D and Summerscales J, Composites 11 (1980) 33

5 Manders P.W and Bader M.G, J. Mats Sci 16 (1981) 2233

Eric Taylor gained his BSc in Chemistry and Polymer Sciences at Loughborough University and now specialises in reinforced composites. Originally with Ever Ready and Dunlop GRG, he joined Fothergill Engineered Fabrics in 1979 and is responsible for R & D. He has presented a number of papers on composites development and applications.

John Weaver graduated in Metallurgy at Nottingham University where he subsequently gained a PhD degree for work on carbon fibre metal matrix composites. With Dunlop Ltd Aviaton Division he was involved until 1977 on the early R & D and production work on carbon/carbon composites for high temperture applications, particularly aircraft brakes. In 1984 he joined Fothergill Engineering Fabrics, a subsidiary of Fothergill & Harvey plc where he is responsible for product development.